Dialogues on Ethical

After lives filled with deep suffering, 74 billion animals are slaughtered worldwide every year on factory farms. Is it wrong to buy the products of this industry?

In this book, two college students – a meat-eater and an ethical vegetarian – discuss this question in a series of dialogues conducted over four days. The issues they cover include: how intelligence affects the badness of pain, whether consumers are responsible for the practices of an industry, how individual choices affect an industry, whether farm animals are better off living on factory farms than not existing at all, whether meat-eating is natural, whether morality protects those who cannot understand morality, whether morality protects those who are not members of society, whether humans alone possess souls, whether different creatures have different degrees of consciousness, why extreme animal welfare positions "sound crazy," and the role of empathy in moral judgment.

The two students go on to discuss the vegan life, why people who accept the arguments in favor of veganism often fail to change their behavior, and how vegans should interact with non-vegans.

A foreword, by Peter Singer, introduces and provides context for the dialogues, and a final annotated bibliography offers a list of sources related to the discussion. It offers abstracts of the most important books and articles related to the ethics of vegetarianism and veganism.

Key Features:

* Thoroughly reviews the common arguments on both sides of the debate.
* Dialogue format provides the most engaging way of introducing the issues.
* Written in clear, conversational prose for a popular audience.
* Offers new insights into the psychology of our dietary choices and our responsibility for influencing others.

Michael Huemer is Professor of Philosophy at the University of Colorado, Boulder. He is the author of more than 70 academic articles in ethics, metaphysics, political philosophy, and epistemology, as well as five other books: *Skepticism and the Veil of Perception* (2001), *Ethical Intuitionism* (2005), *The Problem of Political Authority* (2012), *Approaching Infinity* (2016), and *Paradox Lost* (2018).

Dialogues on Ethical Vegetarianism

Michael Huemer

NEW YORK AND LONDON

First published 2019
by Routledge
52 Vanderbilt Avenue, New York, NY 10017

and by Routledge
2 Park Square, Milton Park, Abingdon, Oxon, OX14 4RN

Routledge is an imprint of the Taylor & Francis Group, an informa business

Library of Congress Cataloging-in-Publication Data
A catalog record for this title has been requested

ISBN: 978-1-138-32828-0 (hbk)
ISBN: 978-1-138-32829-7 (pbk)
ISBN: 978-1-138-32830-3 (ebk)

Typeset in Sabon
by Swales & Willis Ltd, Exeter, Devon, UK

Contents

Foreword

When I became a vegetarian, in 1971, it seemed that everyone I ate a meal with was asking me why I didn't eat meat. My experience was different from that of Michael Huemer, who tells us that people rarely challenged him on his dietary views. Perhaps that reflects the fact that encountering a vegetarian was more unusual in 1971 than it was when he stopped eating meat in the 1980s.

Some vegetarians and vegans think of their dietary choice as a private matter, like one's religious beliefs. That was never my view. I became a vegetarian because I learned about the way animals are treated in factory farms. After some research and reflection, I reached a shocking but inescapable conclusion: there is an atrocity going on, unseen by most of us because it takes place in huge windowless sheds. In each of these sheds there are hundreds, thousands, or – most often – tens of thousands of animals living miserable lives so that we can satisfy our preference for a particular kind of meal.

I knew that I should not support that atrocity by purchasing and consuming its products. Keeping my own hands – or stomach – clean, however, wasn't the point. If we become aware of an atrocity, it is not enough to refuse to take part in it. We should not be bystanders either. We should do what we can to stop it, and telling others about it is an important step in that direction. So although I did not want to push my views on people who showed no interest in them, if someone asked me why I wasn't eating meat, I welcomed the opportunity to tell them about the strong ethical reasons for that choice.

Well, for the first couple of years, I welcomed the opportunity. Then I became tired of repeating the same facts and reasons over and over again, and I thought how good it would be if, whenever someone asked me why I don't eat meat, I could just give them a book and say "Read this." When *Animal Liberation* was published in 1975, I started doing just that.

Now, more than forty years later, I have a better understanding of the many reasons and arguments that people can devise to defend their preference for eating meat. Some of these, like "Animals eat other animals, so why shouldn't we eat them?" have been refuted often before, and yet, like a jack-in-the-box, they keep popping up again. Others are new. In 1971, no one talked about climate change, so very few people had any idea that methane emissions from farm animals were making a major contribution to warming our planet. Although Huemer has chosen to focus his book on the ethics of eating meat from the perspective of what happens to the animals before they become meat, the need to minimize our contributions to climate change is also an important ethical reason for not eating meat.

Philosophical dialogues go back to Plato, who described, or imagined, conversations Socrates had with Athenians about how we ought to live. *Dialogues on Ethical Vegetarianism* is a worthy addition to this evergreen tradition. Lively and easy to read, the dialogues that follow capture the experience of doing philosophy as an active participant and not merely a passive spectator. They are also uncannily accurate in presenting what people actually say. Many passages recapitulate, sometimes almost word for word, what people have said to me in conversations about eating meat.

Huemer makes no secret of what side he is on, but he is first and foremost a philosopher and not a polemicist. Hence he is not interested in scoring cheap points against a weak opponent. To refute a philosophical argument, one must first state it in its strongest form. That is what this book does. Then it shows why those arguments are flawed or do not justify eating meat.

In the future, when people ask me why I don't eat meat, I will tell them to read this book.

Peter Singer
Princeton, New Jersey

Preface

My Journey

The worst thing I have ever done in my life is that I ate meat and other animal products for many years. I can't explain why I did this, except that I simply had reflected little, if at all, on the ethical status of it. I reflected about many things as a child, but not that. Meat tasted good, and no one seemed to see any problem with it. No one brought any issue to my attention. No one asked me to defend my behavior.

It was partly, also, that I knew much less than I now know about the process. I knew what meat *was*; I knew that animals were killed so that I could eat them. I also knew that they were raised in artificial conditions, unable to live the sort of life natural to them. What I did not know was the sheer magnitude of the suffering involved – of that, I would only learn years later, some time in my thirties.

I started thinking about ethical vegetarianism close to thirty years ago, when I was a college student. A friend prompted me to reflect on the morality of our eating habits. Could it be right to raise other conscious beings solely so that we might kill them, cut off parts of their bodies, and chew on them for our own momentary sensory pleasure? I did not read any sophisticated philosophical treatment of the subject; if I had, I might have reached my present views more quickly. Instead, I simply contemplated that question, for how many months I am not sure. But I could not convince myself that this was alright. Once you *thought* about it, it really seemed, on its face, very wrong.

Then, I was still ignorant and confused. Thus, I was not ready to become *vegan* (and there were even fewer vegan advocates then than there are now; I don't think I'd ever met one). I decided that it was wrong to buy meat from farms . . . but for several more years, I continued to buy eggs, dairy, and wild-caught fish. (I would also have accepted other wild-caught meat, if such had been available.) Of course, this made things much easier than a strict vegan ethic would have. And this was not a completely arbitrary position – raising animals in artificial and inhumane conditions in order to kill them is *worse* than killing animals that have lived a natural life in the wild and perhaps also worse than raising animals in order to take their eggs or milk. (Most wild animals, I thought, would likely die even more painful deaths at the hands of predators if not killed by us.)

As I now recognize, that was not a reasonable position; one who opposes cruelty should at least renounce all products from factory farms. (The full case is set out in the following dialogues.) Now, *perhaps* meat is worse than dairy or eggs, but just because A is worse than B does not mean that B is acceptable. There is no need for us to support *any* cruelty.

But, again, these points somehow eluded me for some time. It took me perhaps another ten years to complete the journey to veganism.[1] It is almost unbelievable to me now that the process could have taken so long. My past self now appears very stupid on this point.

But, again, it was not so much a matter of stupidity as of *unreflectiveness*: during those years, I virtually never thought about the permissibility of buying eggs and dairy products. No one made me do so. When eating in a restaurant, for example, no one so much as suggested that there might be an ethical issue with ordering an omelet. If ever anyone challenged my dietary ethics, it was always a meat-eater challenging my refusal to eat *more* animal products. I think that this had an emotional anchoring effect: implicitly, I compared my behavior to the standard behavior of the rest of my society and of my earlier self. Since

1 Or ostroveganism, as discussed in dialogue 4 below.

I was doing better than my past self and almost everyone else around me, I felt little impulse for self-criticism. Of course, this was wrong: the fact that others are doing worse does not make our own behavior acceptable.

I rarely challenged others on their dietary ethics, and they rarely challenged me. I do not like to accuse others of immorality; it feels rude and aggressive, even if those others are in fact behaving immorally. So it happens that we all look the other way, and we all carry on doing what we want without having to think about the morality of it.

I am telling you all this to highlight points of human psychology. I am sure I am not alone in these flaws. I am sure that there are many others who, if not directly confronted about their behavior, will go on doing what they wish with little reflection; many who will rest content with merely being *less* wrong than most of their society; many who will avoid raising moral issues out of a desire to be agreeable. Perhaps, reader, you may even find these tendencies in your own self.

There are other obstacles to becoming vegan that I believe I have observed in others. Most who are confronted with the issue deploy some form of distraction – shifting attention to another issue in the vicinity without directly confronting the morality of their own behavior. Examples: the shift to abstract theory (are there really such things as "moral obligations"? where do they come from? how do we know?); the shift to discussing *my* behavior (as though, if meat-eaters can convict me of hypocrisy by finding some animal product that I buy, then it will be okay for them to continue their own behavior); the shift to hypothetical scenarios in which meat-eating might be acceptable; the choice to focus on only the most extreme animal-rights positions (as though, if you can reject the most extreme possible opposing view, then it's okay for you to continue doing exactly what you're currently doing). These distractions enable one to avoid focusing on one's own current, actual behavior and thus to continue that behavior with a minimum of discomfort.

Aims and Audience

The following dialogues are my attempt to accelerate the process of becoming vegan for others. If you have not yet thought through the issue, I hope this short book helps you to do so. If I could, I would go back in time and give it to my earlier self, when I was ten years old, or whenever I was first able to understand it. Since I cannot do that, I hope to get it into the hands of others who have not yet heard the arguments. I hope you, the reader, will avoid self-distraction and take the opportunity to reflect ethically on your own choices.

I am an academic philosopher, but this is not intended mainly for academics. It is intended for anyone concerned about the ethics of what we eat. I assume no background; the dialogues are meant to introduce the ideas for those not familiar with the existing literature.

These dialogues are meant as, and I believe they are, a *fair* introduction to the issues. In particular, none of the ideas or arguments that appear herein – even those that various readers will find highly implausible – are straw men. Every idea and argument is either something that I myself believe in, or something that I have seen others assert in person or in print. I have sought to present the other side's views, as far as possible, in the way that they themselves would present those views.

But why this book? There are already many academic articles and many books, both academic and popular, on the subject of vegetarianism. The arguments have been clearly and thoroughly set out by other writers. So why set out the case for vegetarianism yet again?

The answer is that I hope to reach a wider audience, or at least a different audience, than those other works – I hope to reach readers who do not read academic journals (which, in all honesty, can be very dry and difficult reading), and who may not wish to read a long, traditional philosophical treatise. The dialogue format is the most accessible and entertaining form for conveying philosophical ideas. And if ever there were an idea that should be made accessible, it is the idea of ethical vegetarianism. That is why I have written this as I have.

These dialogues were first published in *Between the Species*, an open-access, academic journal devoted to animal ethics.[2] I have decided to publish them also in book form, in an effort to reach additional readers. For this version, I have revised the text extensively, added this preface, and added some topic headings to the table of contents, with corresponding notes in the margins of the text, to make it easier for readers to find where a topic is discussed. This edition also includes a new foreword by Peter Singer, the most prominent voice in the animal liberation movement.

All of my profits from the book will be donated to efficient animal-welfare charities, as identified by Animal Charity Evaluators (https://animalcharityevaluators.org/).

2 *Between the Species* 22 (2018): 20–135.

Acknowledgements

I would like to thank Stuart Rachels and Tristram MacPherson for their very thoughtful and helpful comments on the manuscript, as well as two anonymous reviewers for *Between the Species*, and my numerous Facebook friends who discussed early versions of the dialogues. I thank Cheryl Abbate for her excellent work preparing the annotated bibliography.

Michael Huemer
Denver, Colorado, 2018

Acknowledgements

Day 1 Suffering, Intelligence, and the Risk Argument

Setting: Two students, M and V, have met for lunch at their local Native Foods Cafe.[1]

M: Hey, V. I've never been to this restaurant before. Looks nice.

V: Yeah, I come here a lot. It's one of the few vegetarian restaurants in town.

M: (*disappointed*) Oh.

V: Oh, what?

M: Nothing . . . So we're going to be eating sticks and leaves then.

V: No, no. I think you'll be surprised at how good the food is.

M: (*skeptical*) If you say so.

M and V order and then sit down at a table in the corner.

M: So . . . you're a vegetarian.

V: Yep. Been vegetarian for the last three years.

M: Wow. I didn't know you were such a crazy extremist.

V: (*laughs*) Some people would say that. I think it's just the reasonable position.

1 A popular US vegetarian restaurant chain (http://www.nativefoods.com/).

M: Did you know that Adolf Hitler was a vegetarian?[2]

V: (*sigh*) Godwin's law already?[3] Yes, I know. Gandhi was also a vegetarian.

(a) The ethical vegetarian position

M: Well, I guess you can find both good and evil people who have done most things. So what made you give up meat?

V: I figured out that meat-eating is morally wrong.[4]

M: So if you were stranded on a lifeboat, about to die of starvation, and there was nothing to eat except a chicken, would you eat it?

V: Of course.

M: Aha! So you don't really think meat-eating is wrong.

V: When I say something is wrong, I don't mean it's wrong in every conceivable circumstance. After all, just about anything is okay in *some* possible circumstance. I just mean that it is wrong in the typical circumstances we are actually in.[5]

M: So you think it's wrong to eat meat in the circumstances we normally actually face.

2 See Wikipedia, "Adolf Hitler and Vegetarianism," https://en.wikipedia.org/wiki/Adolf_Hitler_and_vegetarianism.

3 Godwin's law: As an internet discussion grows longer, the probability of a comparison involving Hitler approaches 1. Also applies to some non-internet discussions.

4 Here, V follows the work of many philosophers. See, e.g., Peter Singer, *Animal Liberation* (New York, NY: HarperCollins, 2009); Tom Regan, *The Case for Animal Rights*, 2nd ed. (Berkeley, CA: University of California Press, 2004); Alastair Norcross, "Puppies, Pigs, and People," *Philosophical Perspectives* 18, *Ethics* (2004): 229–45, available at http://faculty.smu.edu/jkazez/animal%20rights/norcross-4.pdf; Mylan Engel, "The Commonsense Case for Ethical Vegetarianism," *Between the Species* 19 (2016): 2–31, available at http://digitalcommons.calpoly.edu/bts/vol19/iss1/1/; Stuart Rachels, "Vegetarianism," pp. 877–905 in *The Oxford Handbook of Animal Ethics* (Oxford University Press, 2011), available at www.jamesrachels.org/stuart/veg.pdf.

5 Compare Engel, "Commonsense Case," *op. cit.*, pp. 6–7.

V: Right.

M and V's food arrives, and they start in on two appetizing vegan meals.

M: Okay, you're right: this is better than I thought it would be. I could enjoy coming here once in a while for something different. But it still seems to me like this ethical vegetarianism of yours is an extreme view.

(b) For vegetarianism: the argument from pain and suffering

V: I don't think it's that extreme. Would you agree that pain and suffering are bad?

M: No, I think pain is necessary. You know, there is a rare medical condition in which people are unable to feel pain.[6] As a result, they don't notice when they injure themselves, so they're in danger of bleeding out, injuring themselves further, and so on. It's really quite bad. So you see, pain is actually good.

V: Sounds like you're just saying that given certain conditions, pain can be *instrumentally* good. You don't think it's *intrinsically* good, do you?

M: What do you mean by that?

V: Well, you're just saying pain can sometimes have good *effects*. You're not saying it's good *in itself*, are you? I mean, what if you have to have a tooth drilled at the dentist's office – do you take the anesthetic?

M: Of course. I don't want *gratuitous* pain. I only want pain when it helps me avoid injuring myself, teaches me valuable lessons, or something like that.

V: Agreed. And the same is true about *suffering*, right?

6 Congenital analgesia.

M: What's the difference between "pain" and "suffering"?

V: "Suffering" is broader. Suffering would include things like the experience of being confined in a tiny cell so that you can barely move for a prolonged period of time. That might not exactly be *painful*, but it's certainly a negative experience.

M: Of course, negative experiences are bad, provided that they don't produce some benefit, like teaching you valuable lessons or enabling you to avoid injury.

V: Yes, let's take that as understood. Now, if suffering is bad, in the sense we've just described, then I suppose that *larger amounts* of suffering are worse? Like, if you have intense suffering, for a long period, that's worse than a milder, shorter suffering. Right?

M: Other things being equal, sure.

V: Okay. It also seems to me that it's wrong to cause a very large amount of something bad, for the sake of some minor good. Would you agree with that?

M: That sounds generally right, but I'm not sure that's *always* true . . . What if I make a great sacrifice of my own welfare, to achieve a small benefit for my kids? That wouldn't be wrong; that would be noble. Perhaps foolish, but noble anyway.

V: Okay, I overgeneralized. Let's be more specific. I think it's wrong to knowingly inflict a great deal of pain and suffering on others, just for the sake of getting relatively minor benefits for yourself. That's not some crazy extremist view, is it?

M: (*laughs*) No, it's not.

V: Okay, well, the meat industry inflicts a great deal of pain and suffering on animals, for the sake of comparatively minor benefits. It follows that the meat industry is doing something wrong.

M: What pain and suffering are you talking about?

V: For instance, chickens and pigs are commonly confined in tiny cages where they can't move for their entire lives. Cows are branded with hot irons, to produce third-degree burns on their skin. People cut off pigs' tails without anesthetic. They cut off the ends of chickens' beaks, again without anesthetic. These tails and beaks are sensitive tissue, so it probably feels something like having a finger chopped off.[7]

M: Okay, that's enough. I don't want to hear any more.

V: Why not?

M: It's unpleasant, and I'm trying to eat.

V: But you don't have any problem with eating the products that come from these practices?

M: As long as I don't have to think about it or watch it, I'm okay with it.

V: Isn't that a little hypocritical? If you feel disgusted or horrified by the practices on factory farms, doesn't that suggest that you shouldn't buy their products either?

M: Not at all. There are lots of things that I wouldn't want to look at that are perfectly good and valuable services. For instance, I wouldn't want to watch someone performing surgery, because I can't stand the sight of blood. But that doesn't mean I think surgery is wrong.

7 For more on these and many other forms of cruelty, see Rachels, "Vegetarianism," *op. cit.*, section 1.

V: So the negative reaction you have to animal cruelty is like the negative reaction you have to watching a surgery.

M: Exactly.

V: So it's not because it makes you feel guilty, or because it seems wrong when you see it?

M: Nope. It's just unpleasant to look at, but it's necessary to produce a greater good. The good of human gustatory pleasure.

V: But didn't we agree that it's wrong to inflict great pain and suffering on others, for the sake of relatively minor benefits?

(c) For meat-eating: the argument from intelligence

M: Oh, when I agreed that it was wrong to inflict gratuitous pain and suffering, I thought we were talking about *humans*. Of course it's wrong to inflict gratuitous suffering on other humans. But animals are another matter entirely.

V: Why is that?

M: Oh, that's easy. You see, humans have *intelligence*, and nonhuman animals don't.

V: Do you mean that all animals have zero intelligence?

M: Well, they have drastically lower intelligence than humans.

V: I see.

M: Drastically.

V: And you believe it's morally okay to inflict severe pain on those who are much less intelligent, for the sake of small benefits to those who are more intelligent?

M: Right.

V: And why do you think that?

M: It seems right to me.

V: That isn't obvious to me. Do you have any further reason to give? Something that might make it seem right to me too?

M: Nope. It's just an intuition. It's like "1 + 1 = 2" and "The shortest path between two points is a straight line." You know, self-evident.[8]

V: And you don't have any explanation for why it would be true either? Why intelligence gives one the right to torture the less intelligent?

M: No explanation. It's just a brute fact.

V: I hear that Albert Einstein was really smart. Much smarter than the rest of us, in fact.

M: Yeah, so?

V: So I'm just wondering if that means he would have been justified in torturing us, if he could have gotten some minor benefit by doing so. You know, because he was so much smarter than us.

M: No, he wasn't sufficiently smarter than us. The gap between animals and the average human is much bigger than the gap between an average human and Einstein.

V: Superintelligent aliens, then.

M: What?

V: Say a race of superintelligent aliens lands on Earth. They're way smarter than any human. So can they eat us, torture us for fun, and so on?

M: That doesn't seem right. I'm going to say that there's a threshold level of intelligence that you have to have. If you're above the threshold, then no one gets to torture you. If you're below it – open season.

8 Here, M follows the views of Bryan Caplan, "Reply to Huemer on Ethical Treatment of Animals (including Bugs)," Econlog, October 11, 2016, http://econlog.econlib.org/archives/2016/10/reply_to_huemer.html; "Further Reply to Huemer on the Ethical Treatment of Animals," Econlog, October 14, 2016, http://econlog.econlib.org/archives/2016/10/further_reply_t_1.html; and Richard Posner, "Animal Rights" (debate between Peter Singer & Richard Posner), *Slate*, June 2001, available at https://www.utilitarian.net/singer/interviews-debates/200106--.htm.

V: And what is this threshold?

M: What, you mean like, the actual IQ number?

V: Yeah.

M: Geez, I don't know.

V: Then how can you know that humans are above the threshold and animals are below?

M: Well, I don't know the *exact number*. But I know that it's pretty high. I know it's higher than the level of an animal.

V: You mean it's higher than the intelligence level of even the smartest animal in the world? Higher than that chimp that learned dozens of words of sign language, or the octopus in New Zealand that escaped from his aquarium recently, or the three dolphins that rescued a woman at sea in 1971 . . . ?[9]

M: Oh, I'm not saying that. I'm just saying the threshold for having moral status is above the intelligence level of cows, chickens, or pigs. Or any other animal that we commonly eat.

V: I see. And how do we know that? Assuming there is such a threshold, how do you know that the threshold isn't lower?

M: Again, intuition.

V: Are you sure you're not just saying what is convenient for you, and declaring that to be intuitive?

9 For more, see the following: On Washoe the chimp: the "Friends of Washoe" web site, http://www.friendsofwashoe.org/meet/washoe.html; on Inky the octopus: Wajeeha Malik, "Inky's Daring Escape Shows How Smart Octopuses Are," *National Geographic*, April 14, 2016, https://news.nationalgeographic.com/2016/04/160414-inky-octopus-escapes-intelligence/; on the dolphin rescuers: Annie B. Bond, "3 Stories of Dolphin Saviors," Care2, August 25, 2008, http://www.care2.com/greenliving/3-stories-of-dolphin-saviors.html.

M: Yep, I'm sure.

V: You didn't think about that very long.

M: Sorry, I'm being too cavalier. Let me think about it. (*pauses for three seconds, furrows brow*) Okay, I just introspected very carefully, and I assure you that I'm not being at all influenced by self-interest. By thinking about it purely intellectually, I just directly see that none of the animals whose flesh I like to eat have any moral status.

V: This isn't obvious to me.

M: I guess you're not as morally sensitive as the rest of us are. Don't feel bad. We won't force you to eat meat if you don't want to.

V: Just to clarify, are you saying that because animals have low intelligence, their pain *isn't bad*? Or are you saying that even though their pain is bad, it's still okay to cause a lot of badness for the sake of a trivial benefit?

M: I guess I'm saying the first thing.

V: Their pain isn't bad at all?

M: Well, it's much *less bad* than human pain.

V: How much less? Like half as bad maybe?

M: Oh no, it's at least a thousand times less bad.

V: So the interests of smart humans matter a thousand times more than the interests of stupid animals?

M: Right.

V: I see. Well, even if that's true, factory farming is still wrong.

M: Why do you say that?

V: Because the harm we're causing to the animals is more than a thousand times greater than the benefit we get.

(d) The total amount of pain caused by factory farming

M: Whoa, how do you figure that?

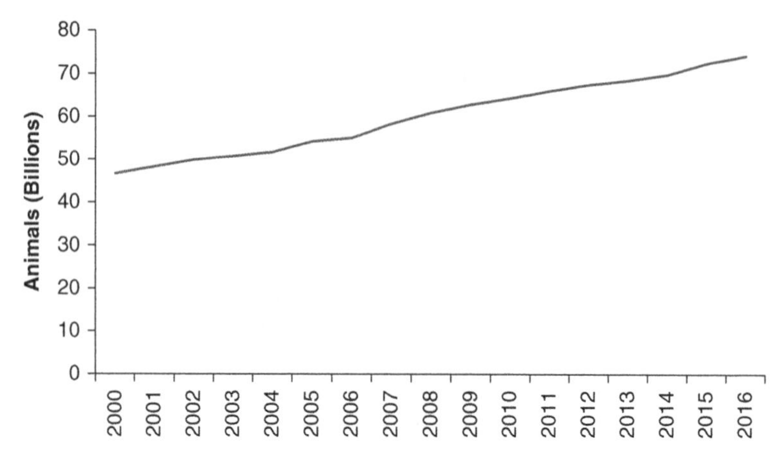

Land animals killed for food, world, 2000–2016

V: It's the sheer numbers. Humans slaughter about 74 billion animals per year, worldwide.[10] And that's just land animals; the numbers of marine animals are much larger.

M: Wow, 74 billion. Well, there are seven and a half billion people to feed.

V: Yes. So the number of animals killed for food in one year is nearly ten times the entire human population of the world.

M: So the average person on Earth eats about ten complete land animals per year . . . Actually, that number sounds low.

V: It's higher in rich countries. For Americans, the number is 31.[11]

M: Okay, that sounds like a lot. But that's just the number of killings. You were complaining about all this pain and suffering. That doesn't tell us how much *suffering* occurs.

10 Source: UN Food and Agriculture Organization, http://www.fao.org/faostat/en/#data/QL, accessed September 5, 2018 (summing totals for Beef/buffalo, Poultry, Sheep/goat, Ass, Camel, Game, Horse, Mule, Meat (not elsewhere specified), Other Camelids, Other Rodents, Pig, and Rabbit).

11 Ryan Geiss, "How Many Animals Do We Eat?" http://www.geisswerks.com/about_animals.html, accessed November 22, 2017.

V: Right, we don't have statistics on the quantity of suffering, since there's no established way of measuring suffering. But almost all of the 74 billion land animals were on factory farms,[12] where the practices are – well, that stuff that you didn't want me to talk about while you're eating. If you imagine the life of a factory farm animal, it seems worse than the life of almost any person – except maybe people who live in torture chambers.

M: But this is only animal suffering. Animals aren't capable of the same level of suffering as humans. They can't be suffering as much as a human would be in those conditions.

V: No? Why do you think that?

M: Because of our intelligence – we're able to have more complex emotions, like shame and humiliation, dread about the further future, and so on. An animal doesn't have an understanding of dignity, and it just lives in the moment.

V: But then, by the same token, human intelligence enables us to comfort ourselves in ways that animals could not. We can take comfort in religion, distract ourselves by imagining better times, and so on. For mistreated animals, their current suffering fills their entire horizon.

M: Wait, are you saying that animals actually suffer *more* than humans from being physically mistreated?

V: I don't know. There are reasons why they might suffer less, and reasons why they might suffer more. Overall, it seems fair to suppose that the suffering of physically abused animals is roughly comparable to that of physically abused people.

12 Nil Zacharias, "It's Time to End Factory Farming," *Huffington Post*, October 19, 2011, https://www.huffingtonpost.com/nil-zacharias/its-time-to-end-factory-f_b_1018840.html.

And so the harm caused by the meat industry is comparable to physically abusing 74 billion people every year.

M: Well, that's ridiculous. Obviously, human and animal interests aren't equally important.

V: Are you sure of that? Some respected ethicists think they are.[13]

(e) Biases of vegetarians and meat-eaters

M: Well, my intuitions are better than theirs. They're probably just soft-hearted animal-lovers who are biased because of their love for animals.[14]

V: They might say that you're biased because of your self-interest.

M: Yeah, but we discussed that, remember? I introspected, and I observed that I wasn't biased. So it's got to be the animal-welfare people who are biased.

V: I remember. So, as you were saying, animal interests are only one thousandth as important as human interests?

M: That's right.

V: Okay, so it's really only one thousandth as bad as torturing 74 billion people a year? So then the meat industry is only as bad as having 74 *million* people tortured per year?

M: Wait, did I say a thousand? I meant a million. Human interests are a *million* times more important than animal interests.[15]

V: Where did you get the "one million" figure from?

13 See Peter Singer, "Equality for Animals?" pp. 55–82 in *Practical Ethics* (Cambridge, UK: Cambridge University Press, 1993), available at https://www.utilitarian.net/singer/by/1979----.htm.

14 Loren Lomasky entertains similar thoughts in "Is It Wrong to Eat Animals?" *Social Philosophy & Policy* 30 (2013): 177–200, at p. 186.

15 Compare Caplan, "Further Reply," *op. cit.*

M: Intuition.[16] I just thought about it, and it seemed obvious.

V: Are you sure you're not just picking numbers for convenience? Like, just saying whatever is required to justify your current practices?

M: Yep. Wait. (*pauses for three seconds*) Okay, I thought about it. I didn't notice any bias on my part.

(f) Intelligence and the badness of pain

V: Let me try a thought experiment on you. Two people have headaches, both equally severe. Assume there are no relevant effects of these headaches beyond the pain. You have one pain reliever that you can give to one of them. You can't split it.

M: Sounds like I should flip a coin.

V: Wait, there's more. One of the people scored higher on the SAT than the other, he's better at solving differential equations, and he has a bigger vocabulary. True or false: you'd better give the aspirin to the smarter person, because his pain is worse, because he's smart?[17]

M: I don't see the point of this thought experiment.

V: I'm just trying to determine whether it's really intuitively obvious that one's intelligence affects how bad one's pain is.

M: Of course, in *that* case it doesn't. Remember, there is a crucial threshold level of intelligence. If you're below the threshold, then your suffering doesn't matter, or only barely matters. When you cross the threshold, your suffering suddenly

16 Some philosophers hold that ethical knowledge derives from "ethical intuition," a kind of direct intellectual awareness of the nature of good, bad, right, and wrong; see Michael Huemer, *Ethical Intuitionism* (New York, NY: Palgrave Macmillan, 2005). For instance, we can see intuitively that pain is bad, or that causing needless harm tends to be wrong. M is seeking to use this theory against V by claiming to have ethical intuitions that support meat-eating.

17 I owe this example to David Barnett.

matters a million times more, even for qualitatively indistinguishable suffering. But after you pass the threshold, no further increases in intelligence make any difference at all; then everyone matters equally.

V: All this sounds arbitrary. Why would all that be true?

M: No explanation is needed. These are just fundamental facts about the moral landscape.

V: And it happens that all humans are above the threshold, so all humans are equal?

M: Yes.

V: So let's say you saw a couple of boys pour gasoline on a cat, then light the cat on fire, just for the fun of watching it writhe in agony.[18] They laugh, showing that they got some enjoyment out of it. To you, this seems perfectly alright?

M: No, sadistic pleasure is always bad.

V: If animal suffering doesn't matter, then what's wrong with taking pleasure in it? Isn't it just like taking pleasure in any morally neutral thing, like watching the grass grow?

M: No, animal suffering is just a *tiny bit* bad. That's enough to make it wrong to take pleasure in it.

V: What if the boys had some other, minor reason for doing it? Like they want to explore a cave, and they need a torch to see by. Burning the cat is slightly more convenient than finding some non-sentient thing to burn. So then it'd be perfectly alright to burn the cat alive?

M: That does seem kind of messed up.

V: How are cats different from cows?

M: Well, food is a more important purpose than having a torch for cave-exploring.

18 Gilbert Harman gives this as an example of seeing wrongness in *The Nature of Morality: An Introduction to Ethics* (New York, NY: Oxford University Press, 1977), pp. 4–5, 7–8.

V: So the boys can burn the cat to later eat it, but not to light the cave?

M: I don't know, still seems messed up. Maybe I'm just biased in favor of cats because I used to have a cat. Maybe it is really okay to burn the cat.

(g) The case of mentally disabled humans

V: What about mentally disabled people? I once met a person who was so severely mentally handicapped that she couldn't talk.[19] Would it have been okay for me to torture her, if it would have given me some minor benefit?

M: No, that would have upset her family.

V: So it's wrong to inflict great pain on an unintelligent being, if doing so upsets some intelligent beings?

M: That's right.

V: Well, factory farming really upsets *me*. Why doesn't that make it wrong?

M: Because the animals aren't yours. They belong to the farmers. So it doesn't matter if *you're* upset. You can't hurt the mentally disabled person, because she belongs to her family, and they would be upset.

V: So only the *family* would be allowed to torture the mentally disabled person.

M: No, they can't do it either.

V: Why not? On your view, isn't that just like farm owners torturing their animals?

M: Yeah . . . but the mentally disabled person is a human. The animals aren't.

V: But we're trying to figure out what's so special about humans. You told me it was intelligence.

19 Profoundly mentally disabled human beings may have difficulty not only talking, but even eating or moving about on their own. See Traci Pedersen, "Profound Mental Retardation," *Psych Central*, April 13, 2016, http://psychcentral.com/encyclopedia/profound-mental-retardation-2.

So if we find a human who lacks intelligence, then it follows that that human *isn't* special. Right?

M: Well . . . they still belong to the same species as us.[20] You see, when you want to figure out how bad someone's pain is, you don't look at their *own* intelligence level. You have to look at the *typical* intelligence level of the *species* that they belong to. If that level is above the threshold, then their pain is really bad. If it's below the threshold, then their pain barely matters.

V: I don't see why the badness of *my* pain would depend on the intelligence of *other* people.

M: Again, this is just a basic ethical axiom.

V: So what if, in the future, a whole bunch of profoundly mentally disabled people were born? So many that it lowered the average intelligence of the human species to below this intelligence threshold that you keep talking about. Then, on your view, it would become okay to torture me? Because I would then belong to a species with a low average IQ?

M: No. You see, there are actually two principles. One, that it's wrong to torture beings who have a high enough IQ. Two, it's wrong to torture beings who belong to a *species* whose IQ is high enough. But it's okay to torture beings whose IQ is below the threshold *and* belong to a species whose average is also below that threshold. That's the moral principle.

20 Here, M follows the position taken by Carl Cohen in "A Critique of the Alleged Moral Basis for Vegetarianism," pp. 152–66 in *Food for Thought: The Debate over Eating Meat*, edited by Steve Sapontzis (Amherst, NY: Prometheus, 2004) at p. 162; and by Posner in his debate with Singer, *op. cit.*

V: And you're sure all this is correct?
M: Totally.
V: Even though you can't explain why these things would be true?
M: It's enough that it's obvious to me.
V: Even though it's not obvious to me.
M: You're biased.
V: Okay. How certain are you of all this?

(h) The argument from moral uncertainty

M: A hundred percent.
V: I don't think you can be a hundred percent certain.
M: Why not?
V: A hundred percent certainty means that it's the most certain thing possible. So nothing else could ever be more certain than these controversial ethical claims you've made. Suppose that God came down from Heaven and told us that one of the following things is the case: either you're mistaken about animal ethics, or you're mistaken in thinking there's a table in front of you now. (*knocks on the table that M and V are sitting at*) Which would you think is more likely?
M: Okay, I'm more sure of the table.
V: So your ethical claims aren't completely certain.
M: Fine. They're only ninety-nine percent certain.
V: So there's a one percent chance that the animal welfare advocates might be right.
M: Yeah, only 1%.
V: Well, that's enough to make meat-eating wrong. If there's a 1% chance that, say, Peter Singer's views of animal ethics are correct, then you're morally obligated to give up meat.
M: How can that be?
V: It's a matter of the ethics of risk. We agree that it's wrong to perform an action that has extremely bad effects on others, for the sake of minor benefits.

M: Sure.

V: But it needn't be *certain* that your action will have bad effects. An action could be wrong just because it *might* have very bad effects.

M: But we accept risks all the time. I mean, when I drive my car, there's always a chance that I'll cause an accident and injure someone. But it's not wrong to drive the car.

V: But it *is* wrong to drive drunk, right?

M: Yeah.

V: Because that creates *too high* of a risk.

M: Yeah. I just don't think the "risk" that we're wronging animals is that high.

V: Would you agree that the acceptable level of risk a person may undertake depends on how bad the bad outcome would be?

M: What do you mean?

V: Say I want to keep a nuclear bomb in my basement. Every day that I keep the bomb there, let's say, there is a tiny chance that something will accidentally set off the bomb. This chance is much lower than the probability that I will kill someone in a traffic accident while driving my car. And yet, it's okay for me to drive the car, but it's not okay to keep the nuclear bomb in my basement.

M: I agree. No one should have personal nuclear bombs.

V: And that's because the harm of a nuclear bomb accident is much greater than the harm of a traffic accident. If I have a car accident, I might kill someone. But if I accidentally set off the bomb, it'll destroy the entire city. So the acceptable risk level is much lower in the case of the nuke.

M: Sounds reasonable. I would add also that you have good reasons for wanting to drive – like, you need to get to work. But I don't think you have very good reasons for wanting to have the nuclear bomb.

V: Agreed. So whether it's okay to do something that might produce a bad effect depends upon (i) how bad the effect would be, (ii) how likely it is that the bad effect would occur, and (iii) how strong of a reason you have for doing the thing.

M: Sounds right.

V: Now, if Peter Singer is right, then the meat industry is about as bad as a practice that tortured 74 billion people a year would be. If there were such a practice, it would be incredibly bad.

M: Good thing Peter Singer isn't right.

V: But if there is a 1% chance that he's right, then the meat industry is about as wrong as a practice that has a 1% chance of torturing 74 billion people a year. Which is about as wrong as a practice that definitely tortures 740 million people a year.[21]

M: That sounds crazy. 740 million?

V: That's 1% times 74 billion. A thing with a 1% chance of doing the equivalent of harming 74 billion people in some way is *1% as bad* as a thing that harms 74 billion people in that way. Which means it is as bad as harming 740 million people.

M: But it's 99% likely that such an action wouldn't harm anyone – then it would be as bad as an action that harms zero people.

V: Sorry, let me rephrase. You have reason to avoid actions that, from your point of view, *might* cause something bad. The strength of this reason is proportional to (i) the probability that the action will cause something bad, and (ii) the magnitude of

21 Compare Rachels' argument in "Vegetarianism," *op. cit.*, p. 898.

the bad outcome that might occur. So, if there is a 1% chance that Peter Singer is right, then the reason we have for abolishing the meat industry is about *as strong as* the reason that we would have for abolishing a practice that tortured 740 million people a year.

M: What if I say there's no chance that Singer is right?

V: If God came down and told you that either Singer is right or this table doesn't exist . . .

M: Okay fine, there's *some* chance that Singer is right. But it's less than 1%.

V: How low? One in a thousand? Then our reason for abolishing factory farming is "only" as strong as the reason we would have to abolish a practice that tortured *74 million* people a year.

M: Well, maybe it's less than that. Maybe there's only a one in a million chance that he's right.

V: Do you really think so? Singer thinks that the badness of a painful experience is just a matter of how painful it is. This doesn't strike me as absurd. Your defense of meat-eating, on the other hand, requires us to accept all of the following:

1. The badness of a pain depends on intelligence, rather than just depending on what the pain feels like.
2. There is a threshold IQ at which the badness of pain suddenly goes up, rather than increasing gradually with IQ.
3. It goes up by something like a factor of a million, rather than, say, only two or ten or a hundred.
4. After that, the badness of pain levels off, rather than continuing to increase.

5 The threshold is safely above the IQ of any animals that are commonly used for food, but below the level of normal humans.
6 Sometimes, the badness of one's pain depends on the typical intelligence of the *species* to which one belongs, rather than one's own intelligence. But
7 It's not *just* the intelligence of the species that matters; rather, it's the larger of the individual's IQ and the average IQ of his species, that determines how bad the individual's pain is.

You don't have any explanation for why any of these things would be true, nor any argument that they are true, apart from the need to say these things to defend our society's current practices. You say they're grounded in intuition, yet all of them strike me as arbitrary at best, not intuitive at all. It's hard for me to see how all of this leaves us with a 99.9999% probability that you're right and Singer is wrong.

M: You're giving me a hard time. But let me ask you this: if you had to kill either a pig or a person, would you really just flip a coin?

(i) Valuing animal vs. human lives

V: Why can't I just not kill anyone?

M: You're driving, your brakes have failed, and you're going to run over a kid, unless you swerve aside and hit a pig.

V: Hit the pig.

M: What if it was ten pigs?

V: Still hit the pigs.

M: What about a hundred pigs?

V: I don't know. Where are you going with this?

M: Well, at last you've admitted that humans are more important than animals!

V: You mean that human lives are more valuable than animal lives.

M: Isn't that what I said?

V: I was just clarifying. How does this make it okay to torture animals?

M: Human pleasure or pain matters more than animal pleasure or pain. You just admitted it.

V: No, I don't agree with that. I think that what's bad about pain is what it feels like. Therefore, *how* bad a painful experience is is just a matter of how bad it feels. It doesn't depend on how big your vocabulary is, or how fast you can solve equations, or anything else that doesn't have to do with how it feels.

M: But then how can you explain that human lives are more valuable than animal lives?

V: I think there are both simple values and complex values. The simple goods and bads in life are pleasure and pain, or enjoyment and suffering. All sentient beings have these, including humans and pigs.

M: Right. But we have a lot more to our lives besides that. Humans can experience moral virtue, understanding of important abstract truths, friendship, great achievements, and so on.

V: Agreed. That's why typical human lives are more valuable than typical animal lives.

M: I'm glad we agree on that. So that's why it's okay for us to eat them!

V: Whoa, no way. That doesn't follow at all.

M: Why not?

V: Because none of those sophisticated goods are at stake here. It's not like you have to buy meat from factory farms in order to survive – then I would be in agreement with you. In fact, many experts think that a vegetarian diet is *more healthy* than a meat-containing diet.[22] And you don't have to eat meat to attain any of those complex, sophisticated goods, either. You don't have to do it to attain moral virtue, or understanding, or friendship, or any great achievement. You're just doing it for pleasure.

M: But human pleasure is more important than animal pleasure or pain!

V: I don't see why.

M: Because humans are capable of more sophisticated *kinds of pleasure* than animals are. It's for the same reason that we are capable of those complex goods that you were just talking about that animals can't have.

V: Sure, maybe we have some more sophisticated kinds of pleasure. But that's irrelevant, because none of them are what we're talking about. We're talking about the pleasure of tasting animal flesh. *That* pleasure doesn't outweigh the suffering of the animal on the factory farm. Our ability to have some other kinds of pleasures isn't relevant to that.

M: You're messing with my brain, V. I need to go home and rest it.

M and V, having finished their meal, start clearing their places.

V: Okay. Are we still on for next week?

M: Sure, I'll talk to you then.

V: Same place, same time?

M: After everything you've said, I guess I won't suggest a burger joint. Okay, we can meet here again.

22 For a review of the evidence, see Engel, "Commonsense Case," *op. cit.*, pp. 12–17.

Day 2 Other Defenses of Meat Consumption

Setting: M and V have met at Native Foods for another delicious vegan meal.

V: Have you thought more about what we discussed last week, M?

M: Yeah. Let me see if I understand what you're saying.

V: Go ahead.

(a) Recap of the previous day's arguments

M: So, factory farming inflicts extreme pain and suffering on 74 billion animals a year, before killing them. Peter Singer thinks animal suffering is just as important as human suffering. If he's right, then factory farming is like an institution that tortures 74 billion people a year. If there's even a 1% chance that he's right, then it's like an institution that tortures 740 million people a year. Or if animal suffering is even 1% as important as human suffering, then again factory farming is like an institution that tortures 740 million people a year.

V: Right. And if there's only a 1% chance that animal suffering is 1% as important as human suffering, then it's "only" comparable to a practice that tortures 7.4 million people a year, which would still be horrific. And since there aren't even any good arguments that animal interests matter drastically less than human interests, it's hard to claim that there's not even a 1% chance that animal interests are even 1% as important as human interests.

M: My argument was that humans are smarter than animals, and that pain matters much less for a less intelligent creature.

V: But you couldn't give any argument for this or any explanation of why it would be true. Which makes it hard to claim certainty that you're right.

M: (*sigh*) Okay. Let's say factory farming is wrong.

V: So you agree that it should be abolished?

M: I guess so. But that doesn't mean that eating meat is wrong.

V: Why not?

(b) The possibility of humane meat

M: Well, meat *could* be produced humanely, without all this cruelty.[1] Therefore, in principle, there could be a meat industry that was morally okay, in which case it would be okay to eat the meat it produced.

V: But remember what the issue was. When I told you that I think meat-eating is wrong, I explained that I mean that it is wrong in the conditions we normally actually find ourselves in. It doesn't matter if there is some hypothetical possible world in which it would be okay.

(c) Are consumers responsible for meat industry practices?

M: Oh yeah. But just because factory farming is wrong, doesn't mean that I'm wrong for buying meat.

V: You know that almost all meat is from factory farms, right?[2]

M: Yeah, I know. But it's not *my* fault that they're so cruel and inhumane. It's not like *I'm* inflicting

1 Compare Lomasky's comments on veal and foie gras ("Is It Wrong to Eat Animals?" *op. cit.*, p. 195).

2 From Zacharias, "It's Time to End Factory Farming," *op. cit.*: "[F]actory farms raise 99.9 percent of chickens for meat, 97 percent of laying hens, 99 percent of turkeys, 95 percent of pigs, and 78 percent of cattle currently sold in the United States."

the pain and suffering on the animals directly. It's the workers on the farms. I don't see why I should be blamed for what *they* do.

V: Because you're paying them for what they're doing.

M: So what?

V: Usually, if it's wrong to do something, then it's also wrong to pay other people to do it. For instance, murder is wrong; so it's also wrong to hire an assassin. If you hire an assassin to kill someone, you can't say that you're not to blame just because you didn't pull the trigger yourself.

M: Yeah, but that's different. In that case, you're specifically *telling* the assassin to commit a murder. I'm not *telling* meat companies to torture animals, and they don't have to do that. They're just deciding to do it that way, on their own initiative.

V: Okay, different example. You have a friend named "Killian," who happens to be a murderer. One day, you offer Killian $20,000 to get you a new car. Killian *could* carry out this task in a perfectly moral manner. But you know that the way he *will in fact* do it is by murdering some innocent person and stealing their car. You know this because Killian has performed tasks like this for you in the past, he always murders people along the way, and you always pay him for it afterward. You don't specifically tell him to murder anyone; you just know that that's the way he does things. So you tell Killian to get a car, he goes off, kills someone, steals their car, gives it to you, and you pay him $20,000. End of story. Did you act morally in this story?

M: I don't like that story. I would never act that way.

V: Of course not. But let's understand why not.

M: Because I don't want to support murder or theft.

V: Glad to hear it. Moreover, it would be wrong to do so, wouldn't it?

M: It seems wrong.

V: Well, that's like buying factory-farmed meat. You didn't tell them to commit acts of extreme cruelty, but you know that that is how they do things, and you keep paying them for the product.

M: But in your story, when I ask Killian to get me a car, the murder and theft hasn't yet occurred, and I'm going to cause it to occur. In the case of buying meat, the animal has *already* been tortured and killed.

V: True. If you like, we can change the example. Say Killian has a business. He murders people and steals their cars, then sells them. That's where all of his cars come from. You go to Killian to buy a car. You buy a car that has already been stolen, and whose owner is already dead. Does this make it okay to buy cars from Killian?

M: I guess not. But I still think that's different from buying meat.

V: How is it different?

M: Well, Killian is a single person with a small business. So when you buy a car from Killian, that might cause him to go out and kill another person and steal their car, to replenish his stock of cars.

V: Right. And when people buy meat, that might cause the meat companies to raise and slaughter more animals to replenish their stock.

(d) Can one person affect the meat industry?

M: But the meat industry is so large that I don't think it would. They're not going to respond to such a small change as a single person giving up meat.

V: Would they respond if a million people gave up meat?

M: Of course. *Then* they'd obviously reduce their production.

V: What if a thousand people gave up meat? Would the industry reduce their production then?

M: Probably.

V: What would you guess is the minimum number of people that would cause the meat industry to reduce its production, if they gave up meat?

M: I don't know.

V: Just take a guess for the sake of argument. It doesn't matter if your guess is wrong.

M: (*shrugs*) Fine. Maybe if a hundred people gave up meat, then the industry would reduce its production. But I can't make a hundred people do that.

V: Alright. Now, if they reduced their production in response to 100 people giving up meat, then they'd reduce it by about the amount that 100 people eat, right?

M: I guess so.

V: But you know, you and I are not the only ones who are thinking about ethical vegetarianism. Other people in our society periodically give up meat in response to ethical reasons. On our current hypothesis, every time 100 people give up meat, the industry reduces its production by the amount eaten by 100 people.

M: Okay.

V: Well, that means that you might trigger a reduction in production like that. If 99 other people have given up eating meat since the last time they reassessed their production, then you'll be the hundredth. Then you'll push them over the 100-person threshold, causing them to reduce production by the amount that 100 people eat.

M: Yeah, but that seems really unlikely.

V: It is: it's only 1% likely. But if it happens, it'll reduce production by 100 times the amount that a single person eats. So it's worth it.[3]

M: Okay, but this reasoning was just based on a guess that I made. I just *guessed* that it takes 100 people to cause the meat industry to respond.

V: True. But similar reasoning would apply no matter what guess you'd made. If you had guessed "86" instead of "100," then I would have said: there's a 1/86 probability that you'll cause the industry to respond, in which case they'll reduce production by the amount that 86 people eat. It's still worth it to give up meat.

M: I see. But you're still assuming that there was *some* correct answer, some specific number of vegetarians who would induce the meat industry to reduce production.

V: And that's not true?

M: I don't think so.

V: You mean that the industry won't reduce production no matter how many people give up meat?

M: No, I don't mean that. I mean that there might be no *particular number* at which it would happen.

V: If a million people gave up meat, one after another, the meat industry would reduce production at some point, right?

M: Presumably.

V: Well, whenever it happened, it would be after some particular number of people. There's no way

3 Here, V follows the reasoning of Peter Singer ("Utilitarianism and Vegetarianism," *Philosophy & Public Affairs* 9 [1980]: 325–37 at pp. 335–6); Gaverick Matheny ("Expected Utility, Contributory Causation, and Vegetarianism," *Journal of Applied Philosophy* 19 [2002]: 293–7); Norcross ("Puppies, Pigs, and People," *op. cit.*, pp. 232–3); and Rachels ("Vegetarianism," *op. cit.*, p. 886).

of reducing production just in general, without doing it at any particular time.

M: Yeah. But maybe each time someone gives up meat, there's just a *chance* that they'll reduce production, but it's not determined in advance exactly when it will happen.

V: Maybe, but this still doesn't undermine my argument. As long as there's that chance that they'll reduce production, you have a reason to give up buying meat, since you could trigger a large reduction.

(e) How industries respond to reduced demand

M: But maybe if a bunch of people give up meat, instead of reducing production, the meat industry will just lower prices so they can still sell all the meat they produce, to the remaining meat-eaters.[4]

V: Well, you've brought up a real point from economics, but you've oversimplified it. In standard economic theory, when demand for a product declines, the producers lower the price *and* reduce production. So if many people give up eating meat, then they'll lower the price *and* reduce production.

M: Okay. So with the lower price, *other* meat-eaters will increase *their* meat-eating. Which will reduce the impact of my sacrifice. Before I go to all the trouble of converting to a vegetarian diet, I want to know how much it's going to impact meat production.

V: It happens that agricultural economists have looked into this. They've done empirical studies of the market for meat in the US. They find that,

4 Daniel D'Amico makes a similar suggestion in "Don't Prioritize the Welfare of Animals over Humans," *Reason*, October, 2018, https://reason.com/archives/2018/09/26/proposition-libertarians-shoul1.

on average, if you reduce your meat purchases by one pound, producers will decrease their production by 0.68 pounds (for beef), 0.76 pounds (for chicken), or 0.74 pounds (for pork).[5]

M: Wait, you're saying the industry responds to *every* one-pound change in demand? You mean a single pound per year?

V: No, they don't respond every time. I'm saying that's the *average* effect. And because it's an average, it doesn't matter if you're talking about a pound per day, a pound per year, or whatever.

M: Okay. But your argument was assuming that meat production is *falling* because of people becoming vegetarian. I've heard that overall meat consumption *increases* in most years. Not to cast any aspersions on your persuasiveness, but new meat-eaters are entering the market faster than people are giving up meat.[6]

V: True. But all the above arguments also apply in reverse.

M: What do you mean "in reverse"?

V: I mean, just as a decline in demand causes a *drop* in production, an *increase* in demand causes a *rise* in production. So, according to the economists I was referring to, if you buy one more pound of meat, that'll cause the industry, on average, to *increase* their production by 0.68 pounds (for beef), 0.76 pounds (for chicken), or 0.74 pounds (for pork). If meat production is on

5 F. Bailey Norwood and Jayson L. Lusk, *Compassion, by the Pound: The Economics of Farm Animal Welfare* (Oxford, UK: Oxford University Press, 2011), p. 223. The comparable figures are 0.56 for milk, 0.69 for veal, and 0.91 for eggs.

6 See Carrie R. Daniel, Amanda J. Cross, Corinna Koebnick, and Rashmi Sinha, "Trends in Meat Consumption in the United States," *Public Health Nutrition* 14 (2011): 575–83.

the rise, then by giving up meat, you reduce the rate at which it rises. If it's on the decline, then you increase the rate of decline.

(f) Farm animals only exist because of meat consumption

M: Wait a minute. If the meat industry reduces its production, then farm animals won't be better off; there will just be fewer of them. It's better to have a low-quality life than not to live at all. So we're doing future generations of animals a favor by eating animals today![7]

V: Would you accept this argument if it were applied to people? What if a particular race of people were bred solely to serve as slaves? Then you could say that those particular people would not have existed if not for the practice of slavery. Would this make slavery okay?

M: Of course not. But maybe that's because humans have more rights than animals do. Once a person exists, you have to respect their rights, no matter what the reason was for bringing them into existence. But you don't have to respect animals' rights.

V: And why don't animals have rights?

M: I don't know; I haven't figured out the basis for rights yet.

V: Well, it doesn't matter, because life on a factory farm is much worse than no life at all. It would be much better not to create billions of miserable lives.

M: Really? Why do you think it's worse than no life at all?

7 Here, M follows the reasoning of Lomasky ("Is It Wrong to Eat Animals?" *op. cit.*, pp. 190–91), John Zeis ("A Rawlsian Pro-Life Argument Against Vegetarianism," *International Philosophical Quarterly* 53 [2013]: 63–71, at pp. 69–70), and Jan Narveson ("Animal Rights Revisited" pp. 45–69 in *Ethics and Animals*, edited by Harlan B. Miller and William H. Williams [Clifton, NJ: Humana Press, 1983], at p. 55).

V: I could start describing the conditions on factory farms again, but I think you need to see it. Go look up some videos on factory farming, then I think you'll agree with me.[8]

M: Okay. But this all assumes *utilitarianism*, doesn't it? I mean, you're assuming that the morally right action is just a matter of what produces the most expected pleasure, or the least expected pain, where you weight each good or bad possible consequence by the probability that it will occur.[9]

(g) Utilitarian and non-utilitarian reasons against eating meat

V: No, I'm not assuming utilitarianism. I thought *you* were assuming utilitarianism.

M: Me? How so?

V: *You* were making the argument that it's okay to buy products from an immoral industry, provided that doing so doesn't cause them to increase their immoral actions. That's something a utilitarian might say. That's not *my* view.

M: What would your view be?

V: *My* view would be that it's wrong to financially reward extremely immoral businesses, regardless of whether you're *causing* them to do it, or if they've already done it and you're paying them after the fact.

M: If it's not contributing to the amount of immoral behavior, what's wrong with it?

V: Two things: one, you're rewarding wrongful behavior, which is unjust. You're contributing to making it so that immorality pays –

8 See, for example, Mercy for Animals, "What Cody Saw Will Change Your Life" (video), https://www.youtube.com/watch?v=7FhHgYjymNU, accessed November 29, 2017.

9 This is a popular theory among ethicists; for discussion, see J. J. C. Smart and Bernard Williams, *Utilitarianism: For and Against* (Cambridge, UK: Cambridge University Press, 1973).

M: But that's always true. A lot of immoral behavior has been paying for a long time.

V: But from a moral point of view, you're responsible for your own role in the system. You're not necessarily obligated to fix the world's injustices, but you are obligated not to become a part of them. Second, whether or not immorality pays, you have a duty not to become party to a crime after the fact. You should not willingly make it the case that great wrongs are done *for you*.

M: I'm not sure it would be true that animals were tormented and killed "for me"; after all, the people on the farms don't know anything about me in particular.

V: It's done for the meat customers. You're a part of that class, so it's done in part for you, as well as all the other meat customers.

M: But if a wrong is going to be committed regardless of what we do, shouldn't we make the most of it by taking whatever advantage can be gained from it?

V: I doubt you would think that in other contexts. Say you live in Nazi Germany. Someone offers you a job helping load Jews into gas chambers for execution. It pays a little more than your current job. If you turn down the job, they'll just get someone else to do it. Would you take the job?

M: No, but . . .

V: The wrong is going to be done anyway; why not take personal advantage of it?

M: That's kind of messed up.

V: That's my point.

M: So you're assuming an *anti*-utilitarian ethics, then.

V: No, I'm saying that meat-eating is wrong either way. On *any* reasonable moral view, it's wrong to inflict severe pain and suffering on others without good reason. So factory farming is wrong. If you're a utilitarian, you should give up meat because doing so reduces the expected number of animals suffering on factory farms. If you're not a utilitarian, you should give up meat for the additional reason that you don't want to be party to serious immorality.

M: Okay, fine. But all this turns on the pain allegedly suffered by farm animals. How can we really be sure that nonhuman animals feel pain?[10]

(h) Do animals feel pain?

V: You can't be 100% certain of *anything* about the external world. I mean, I can't be *100%* sure that *you* feel pain. Maybe you're just a mindless automaton. Does that mean I should feel free to torture you?

M: No, that won't be okay.

V: Which shows that I don't have to be *100% certain* that I'm inflicting pain in order for my action to be wrong. If there's even a good chance that you can feel pain, I shouldn't torture you.

M: Okay, then why is there even *a good chance* that animals feel pain?

V: One, they act like they feel pain. They sometimes scream, try to escape, and so on. They do this under the same conditions that would make *you* scream or try to escape, e.g., if someone cut off one of your body parts. Two, animals have the same physiological structures that explain *your*

10 René Descartes famously held that nonhuman animals are mere mindless automata. See his letter to Henry More dated February 5, 1649 in *The Philosophical Writings of Descartes*, vol. 3, tr. John Cottingham, Robert Stoothoff, Dugald Murdoch, and Anthony Kenny (Cambridge, UK: Cambridge University Press, 1991), pp. 365–6.

capacity to feel pain – the same sort of pain sensors in the body, connected up to the same brain areas. That's why no animal scientist seriously doubts that farm animals feel pain.

(i) Animals eat each other, so why can't we eat them?

M: Alright, so they feel pain. But animals eat each other all the time. So why shouldn't we eat them?[11]

V: Are you saying that anything done by an animal is morally permissible for you to do?

M: Well, no. If an animal killed a person, that wouldn't show that it was okay to kill people. It's just that the animals *couldn't complain* about being eaten by us, since they eat each other.

V: You know, humans sometimes kill other humans. Would you accept this reasoning: "humans can't complain about being killed, since they sometimes kill each other"?

M: Well, maybe the particular people who have committed a murder can't complain if we kill *them*. We can execute murderers. Of course we can't just kill any human merely because some other humans have murdered.

V: So by similar logic, animals that have killed other animals may be killed in turn? Like capital punishment for animals?

M: Sure.

V: But the chickens, pigs, and cows on the farms don't eat each other, and they don't kill other animals.

11 Here, M follows the reasoning of Benjamin Franklin (*The Autobiography of Benjamin Franklin*, edited by Charles W. Elliot [Pennsylvania State University Press, 2007], p. 32): "[W]hen the fish were opened, I saw smaller fish taken out of their stomachs; then thought I, 'If you eat one another, I don't see why we mayn't eat you.' So I din'd upon cod very heartily, and continued to eat with other people, returning only now and then occasionally to a vegetable diet. So convenient a thing it is to be a reasonable creature, since it enables one to find or make a reason for everything one has a mind to do."

M: Well, chickens sometimes eat insects and worms.

V: Okay, chickens eat other species, so it's okay to kill chickens. But people also eat other species, so . . . it's okay to kill people?

M: That's awful.

V: I'm just following the logic of your argument.

M: Fine. Now let's follow the logic of your argument. You're saying it's wrong to eat animals. Then it must also be wrong for animals to eat other animals, right? Is it wrong for a lion to eat a gazelle?

V: I dunno, what do you think?

M: I say it's *not* wrong, because it's natural for animals to eat each other.

(j) Free will and moral agency

V: Do you think lions are moral agents?

M: What do you mean by "moral agents"?

V: Do they have free will? And are they able to regulate their behavior according to moral principles?

M: No to both. Animals just act on instinct.

V: Then nothing is morally right or wrong for a lion.

M: You're saying morality only applies to things with free will that can regulate their behavior morally?

V: Don't you agree? You don't blame a baby for crying on an airplane, or a hurricane for destroying a city, or a lion for killing a gazelle. Because none of them are capable of regulating their behavior morally.

M: But then how can similar actions be wrong for us?

V: We have free will and are able to regulate our behavior according to moral principles.

M: That's so unfair. Lions get to do whatever they want, but we have to restrain ourselves?

V: That's the nature of morality.

(k) Should we stop animals from killing each other?

M: Okay, lions can't restrain themselves. But do you think *we* should stop lions from killing gazelles?[12]

V: If you can figure out a way of doing that without killing all the lions and disrupting the ecology, then we should consider it. In the meanwhile, though, I know a way that we could prevent *ourselves* from slaughtering animals, without us dying. We could just eat vegetables.

(l) Do rights imply obligations?

M: But if animals can't have moral obligations, then doesn't that mean that they can't have moral rights either? I've heard from conservative moralists that rights imply obligations.

V: My case for vegetarianism didn't rely on any claims about "rights." Remember that it was all compatible with utilitarianism. I'm only assuming that you shouldn't inflict enormous pain and suffering for minor reasons.

M: Let me rephrase. If it's impossible for some creature to do wrong, then it's also impossible for anyone to do wrong *to* that creature.

(m) Does morality protect those who cannot understand morality?

V: Why do you think that?

M: Morality only protects those who can understand morality.[13]

12 Ned Hettinger raises this question as a problem for animal welfare advocates in "Valuing Predation in Rolston's Environmental Ethics: Bambi Lovers versus Tree Huggers," *Environmental Ethics* 16 (1994): 3–20. For replies, see Jennifer Everett, "Environmental Ethics, Animal Welfarism, and the Problem of Predation: A Bambi Lover's Respect for Nature," *Ethics and the Environment* 6 (2001): 42–67; Anne Baril, "Equality, Flourishing, and the Problem of Predation," pp. 81–103 in *The Moral Rights of Animals*, edited by Mylan Engel and Gary Comstock (Lanham, MD: Lexington Books, 2016).

13 This view is taken by Timothy Hsiao ("In Defense of Eating Meat," *Journal of Agricultural and Environmental Ethics* 28 [2015]: 277–91), who claims that infants and the mentally disabled possess a "root capacity" for rationality even if they can never exercise it. Cohen ("A Critique of the Alleged Moral Basis for Vegetarianism," *op. cit.*) argues on similar grounds that animals lack rights, though we may still have weighty obligations toward them.

V: Again, why do you think that?

M: It sounded good when I said it.

V: Let me give you a couple of examples. You're saying morality only protects those who can understand morality. Babies can't understand morality. It follows that, on your view, morality doesn't protect babies. So it would be alright to torture them.

M: That's terrible.

V: Another example. Say you have an adult human who can't understand morality. Like a mentally disabled person. Can we torture them?

M: No.

V: Or a psychopath – again, they can't understand morality. Can we torture them?

M: Well, if they commit crimes, we should put them in jail.

V: Don't change the subject. I'm not asking about jail; I'm asking about torture. And I'm not asking if you can do it because they committed a crime; I'm asking if you can do it simply because *they* can't understand morality.

M: Well, no, that would be wrong.

V: So it's not true that morality only protects those who can understand morality. Morality protects infants, mentally disabled people, and psychopaths from gratuitous infliction of pain and suffering. So why not animals?

(n) The social contract theory of ethics

M: Maybe morality is the result of a social contract, so it only protects those who are members of our society.[14] The mentally disabled person, baby, and psychopath are all members of our society, despite their limited understanding.

14 Jan Narveson takes a similar view in "On a Case for Animal Rights," *The Monist* 70 (1987): 31–49.

V: So if you meet a person from another society, you can torture and kill them?

M: No, because they *could* join our society.

V: What if they can't because our society has rules that permanently exclude them? Then we can torture them?

M: No, because they could be members of our society if we were to change our practices.

V: Okay. So a mentally disabled person from another society is protected because our society could adopt that person as a member?

M: I guess so.

V: Then why wouldn't animals also be protected because our society could adopt *them* as members?

M: Animals are never true members of society, not even when we treat them as if they were.

V: Why not?

M: Well, they never really reciprocate – they don't understand the social rules, they don't engage in public deliberation; things like that.

V: Okay. But the same is true of severely mentally disabled people. So we could torture them, right?

(o) Is meat natural?

M: No. But . . . eating meat is natural! People have done it for all of history. We have teeth adapted for chewing meat, see? (*points to canines*) But eating mentally disabled humans isn't natural.

V: You're a big fan of naturalness, are you?

M: Sure . . . sometimes. Sometimes I like to follow nature.

V: Those hot dogs that you enjoy: do you know how many unnatural ingredients they stick in those things?

M: So maybe I should give up the hot dogs. But other meat products are more natural.

V: What do you mean by "natural"?

M: You know, meat-eating follows our instincts. We evolved to do it. It's what our ancestors have always done.

V: Our ancestors didn't run factory farms.

M: Yeah, but they ate meat, and we need factory farms today in order to provide the amount of meat we want to eat at a reasonable price.

V: Our ancestors also did some other things that I bet you wouldn't approve of. Slavery, wars of conquest, oppression of women, torture . . .

M: Okay, scratch the point about ancestors. But it's still natural because it follows our instincts and we evolved to do it.

V: Do you think that everything that's natural is good?

M: Well, not necessarily *good* per se . . .

V: But at least okay? Is everything that's natural something that's okay?

M: Sure.

V: You know, cancer is natural. So are earthquakes, hurricanes . . . .

M: Okay, those things are bad. But I'm just talking about behavior. *Behaviors* that are natural are okay.

V: I think war is natural too.

M: How could *that* be natural?

V: Well, people have been doing it for all of human history. That's some evidence that it's natural for humans, isn't it? Just as eating meat is natural for us? Primitive tribes make war even more than we do.[15] There seems to be some sort of human instinct to conquer and dominate other people.

15 See Steven Pinker, *The Better Angels of Our Nature* (New York, NY: Viking, 2011).

M: I don't see how an instinct for war could evolve – wars are so destructive.

V: According to one theory, men in primitive tribes would attack a neighboring tribe to kill the neighboring tribe's men and kidnap and rape its women. In our evolutionary past, the men who succeeded in doing this sort of thing tended to leave behind more offspring than peaceful men who stayed at home. They passed on their genes for aggressive behavior. That's how the instinct for war evolved.[16]

M: I don't know that I want to buy into that. That's a very cynical and nasty theory.

V: Yeah, it's unpleasant. But at least it's a possible explanation of war, right?

M: I guess it's possible.

V: So here's my question: would you say that *if* that theory is correct, then war is good?

M: I'm pretty sure that war is bad.

V: But if the theory I described is correct, then war is *natural*. So then it would have to be good, right?

(p) Are animals missing souls?

M: Well, I guess not everything natural is necessarily good. But I've also heard that God gave humans souls, and He didn't give souls to animals.

V: What's a soul?

M: I'm not sure exactly. It's some immaterial component of a person that goes to heaven after you die.

V: And how do we know that there are any such things?

16 See Steven Pinker, *How the Mind Works* (New York, NY: W. W. Norton & Co., 1997), pp. 513–17.

M: Perhaps because we have conscious experiences. The existence of a soul explains how inanimate matter differs from conscious beings.[17] There's a certain way that it feels to be us, but there's nothing that it feels like to be a hunk of matter.

V: In that case, I think animals have souls too. They have experiences; there's something that it feels like to be a cow.

M: But they can't reason using abstract concepts! Also, they don't go to heaven or hell when they die, according to traditional religions.

V: So their souls are non-rational and mortal.

M: Right. Only humans have rational, immortal souls.

V: I'm not seeing how this makes it okay to torture and kill animals for trivial reasons. If anything, now it sounds like it's even worse to kill an animal than a human. At least the human gets an afterlife; the animal is just gone forever.

(q) Does the Bible support meat-eating?

M: But the Bible says that God granted us dominion over the Earth and all its creatures.[18]

V: Did he give us the Earth and its creatures to torment, or to watch over as responsible stewards?

M: Well, "responsible steward" sounds more like the sort of thing God would expect of us.

V: Is there a Bible verse that says torturing animals for minor reasons is okay?

17 See René Descartes' famous defense of mind/body dualism in his *Meditations on First Philosophy*, in *The Philosophical Writings of Descartes*, vol. 2, tr. John Cottingham, Robert Stoothoff, and Dugald Murdoch (Cambridge, UK: Cambridge University Press, 1984).

18 Genesis 1:26 (King James Version): "And God said, Let us make man in our image, after our likeness: and let them have dominion over the fish of the sea, and over the fowl of the air, and over the cattle, and over all the earth, and over every creeping thing that creepeth upon the earth."

M: No, but I can't find any verse that says it's *not* okay.

V: It also doesn't say that insider trading is wrong.[19]

M: What does that have to do with this?

V: My point is that the Bible isn't an exhaustive list of everything that's right and wrong. We also have to exercise our conscience.

M: Okay, I'm not defending factory farming. The point about the soul was just to explain why we have rights and animals don't – in response to the extreme animal rights advocates.

V: I don't think it even explains that. It's at least possible that we don't really have immortal souls, right?

M: I guess it's possible.

V: Okay. *If* it turns out that we don't have immortal souls, will we then have no rights?

M: No.

V: So it looks like the reason you believe we have rights is not really that we have immortal souls.

M: (*sighs, looks at watch*) Okay. I haven't figured out why you're wrong yet, but I'll try again next week.

V: Agreed. Same time and place?

M: Actually, there's a new Chinese restaurant I'd like to try. It's supposed to have a great Kung Pao Chicken.

V: No.

M: Come on, *you* don't have to eat the chicken. They have vegetarian dishes too.

19 Insider trading is a crime wherein individuals buy and sell stocks based on "inside information" not available to the public. For instance, a company executive might buy stock in a company because he knows that his own company is planning to merge with the other company, which will drive up the price. This is prohibited in the US, UK, European Union, and many other countries.

V: I'm not going there to watch you eat meat.

M: Why not?

V: Have you forgotten the conversation we just had?

M: No, I know you're against eating meat. I just didn't realize you had such a problem with *other* people eating meat. That's so *judgy*.

(r) Judging meat-eaters

V: You must not have understood my position. I haven't just been reporting personal preferences; I'm not saying "I personally prefer not to eat meat." I'm saying it's morally wrong. That's what all the arguments have been about. That means it's just as wrong for you to do it.

M: I understand that that's your opinion. But what gives you the right to judge *me*?

V: Are you judging me for judging you?

M: Uh oh, this feels like one of those self-referential paradoxes. If it's wrong to judge people, then it's wrong to judge that it's wrong to judge people . . .

V: Look, it doesn't require any special rights to make a moral evaluation. If you see someone do something, and you have enough evidence that it was wrong, then you can and should draw the conclusion that they acted wrongly. You don't need to be some special authority figure with special rights or anything. If you saw me beat up a child, you would rightly conclude that I was behaving badly. You wouldn't refuse to draw any conclusions, just to avoid being "judgy."

M: Okay fine; I'll meet you here again so I won't offend your delicate sensibilities. But I still want you to know that I'm feeling pretty judged here.

Day 3 Consciousness and Rational Belief

Setting: M and V sharing a third wonderful vegan meal.

M: Hey V, I think I figured out why human pain matters more than animals'.

V: Do tell.

M: Okay, so there are positive and negative mental states, right? Pleasure, happiness, and other forms of enjoyment are positive. Pain, unhappiness, and other kinds of suffering are negative.

V: Sounds right.

(a) The theory of degrees of consciousness

M: But mental states, in general, can be conscious or unconscious. You can have conscious or unconscious desires, beliefs, and even emotions.

V: All this is well known.

M: Here's the interesting part. Consciousness isn't an all-or-nothing thing; it's a matter of degree. Mental states can be *more* or *less* conscious, not simply conscious or unconscious.

V: So you can have a semi-conscious belief or desire, one that you're only half aware of?

M: Right. Now, when it comes to negative states, the less conscious they are, the less bad. If there could be a completely unconscious pain, then it wouldn't be bad at all in itself.

V: Maybe. But I'm not sure the idea of an unconscious pain makes sense. Unconscious belief, sure. But unconscious pain?

M: But there can be *more* or *less* conscious pains. Say you have a back pain. I decide to distract you from it by engaging you with a delightful philosophical paradox. You start to pay less attention to the pain, and so you start to become less conscious of it. After we've been arguing for a half hour, you've forgotten about your pain.

V: Yeah, I've had experiences like that.

M: And my distracting you would be a good thing, right?

V: Fair enough. So you're saying that the pain becomes less *bad* as it becomes less *conscious*. Where are you going with this? You're not going to claim that animal pain is always unconscious, are you?

M: No, but it might be *less* conscious than typical human pain. Animals have much less self-awareness in general than humans. Some people doubt whether animals are self-aware in general. So it's plausible that all their mental states have only a low level of consciousness. They're only dimly aware of the things they are aware of. In that case, their pain would be less bad than typical (fully conscious) human pains.

V: Interesting theory. This is the first time I've heard someone give an explanation for why animal pain matters less than human pain that makes sense. Usually, you guys pick on arbitrary properties, like IQ.

M: So you don't think level of consciousness is morally arbitrary?

V: No, that really seems to matter to the intrinsic value of an experience. This time, your moral claim is actually intuitive.

M: At last, you've conceded that I made a good point!

V: Yes, but let's explore a few implications of this theory. First, say we have a newborn baby . . .

M: Oh no; it's back to the infants and mentally disabled people again?

V: Well, they seem to have similar cognitive capacities to nonhuman animals, so it's good to test our intuitions on them, to make sure we're not influenced by mere bias against other species.

M: Oh, I'm sure I'm not biased against other species. I just have a rational assessment of their capacities.

V: I'm not convinced that infants or severely mentally disabled people have more self-awareness, or more "consciousness" as you say, than animals. So would it be okay to torture infants and mentally disabled people?

M: That doesn't seem right. Maybe their experiences are all still fully conscious, even though they have much lower general intelligence.

V: Maybe. Or maybe not. Do you want to rest the ethical treatment of these people on that speculation?

(b) Erring on the side of caution

M: Hmm. Well, since we're not sure of their level of consciousness, I would say it makes sense to err on the side of caution and not inflict needless suffering on them.

V: That sounds completely reasonable. Similarly, since we're not sure of the level of consciousness of nonhuman animals, it makes sense to err on the side of caution and not inflict needless suffering on *them*.

M: I guess I'm saying that animals are less conscious than human infants, or at least less likely to pass any given threshold level of awareness.

V: Is there any evidence for that?

M: Maybe the fact that infants are *going* to be fully conscious later?

V: They are, but it's also true that they *were* fully nonconscious in the recent past (at an early stage of fetal development). They're in transition from fully unaware to fully aware beings. I don't see why you should assume that, upon emerging from the womb, they're immediately more conscious than an adult animal. Who knows, maybe animals are *more* conscious than infants.

M: I admit, it's hard to say. I can't think right now of a way we could test for degrees of consciousness of someone's mental states.

V: So it looks like we should err on the side of caution and avoid hurting *any* of these beings – animals, infants, or mentally disabled people – unless we have a very good reason.

M: Or maybe we should reason in the opposite direction. Maybe we should say that since we aren't sure of their level of awareness, we should discount the interests of all of these beings. We should give preference to normal, intelligent adults.

V: Maybe. But how much preference? Would you be willing to say that the pain of a normal adult matters a million times more than the pain of an infant?

M: What do you think I am, some kind of crazy extremist?

V: I'll take that as a no. Then you shouldn't take the crazy extremist view about animals either.

M: But do you agree that human pains are more important than animal pains?

V: I don't know, but it doesn't matter. It doesn't matter if a human pain is 50% worse, or twice as bad, or ten times as bad, as a similarly caused animal pain. The amount of animal suffering we're causing each year is still vastly greater than the benefits we gain. Remember, we're torturing and killing *74 billion* land animals per year.

(c) The use of "torture"

M: I remember. But why do you keep using that emotionally charged word, "torture"?

V: It's an accurate description. If a human being were confined in a tiny cage all day, forced to sit in his own excrement, forced to breathe ammonia, with a small part of his body having been cut off, you wouldn't hesitate to call it "torture." Of course the word has negative emotional connotations, because the *phenomenon* that it refers to is awful. That doesn't make it an inaccurate or unfair description.

M: (*sigh*) Okay, so even if my theory about degrees of consciousness is correct, I'm *still* obligated to give up meat.

V: That's about it.

(d) Why prioritize the animal welfare cause?

M: But I don't understand why you're so obsessed with this problem. Shouldn't we first solve the enormous problems our own species faces, before we start worrying about other species?

V: What problems do you mean?

M: You know, like war, poverty, and disease.

V: We can work on all those things while at the same time being vegetarians.

M: Yeah, but you seem to spend a lot more time worrying about the problems with the meat industry than you do worrying about human problems.

V: If you meet an AIDS researcher, would you criticize them for trying to cure AIDS, when they could instead be trying to cure cancer?

M: I guess we can work on both. But cancer and AIDS are of roughly comparable badness.

V: Okay, say you meet someone who's worried about the problem of internet bullying. Would you tell them to stop worrying about bullying until we first cure cancer?

M: I guess not. But you still haven't answered what makes *you* so interested in animal welfare.

V: You're right, I haven't. I think factory farming is the world's greatest problem, and it's also one of the most neglected.

(e) Is factory farming the world's worst problem?

M: How could a problem for mere animals be worse than the worst human problem?

V: It's estimated that the number of people who have *ever* lived on Earth is about 108 billion.[1] So in just two years, the meat industry slaughters more animals than the total number of human beings who have ever existed. Most endure great suffering before the slaughter. It may be that a few years of factory farming causes more suffering than all the suffering in human history.[2]

M: But there are other problems that humans face besides suffering.

V: Granted. But suffering is still a huge problem. If a single cause accounted for *all* the human suffering in history, would that thing be a big problem or a small problem?

M: Big problem.

V: And if that thing was a practice that we could change at the cost of a moderate amount of pleasure for a couple of years, would you think it worthwhile working to change that practice?

M: Sure. But I'm not convinced that factory farming really causes that much suffering.

1 Live Science, "The Dead Outnumber the Living," February 7, 2012, https://www.livescience.com/18336-human-population-dead-living-infographic.html.

2 Compare Rachels' argument that American factory farms over the last twenty years have caused at least five thousand times more suffering than the Holocaust ("Vegetarianism," *op. cit.*, p. 897). But note that Rachels does not take account of the short lifespan of farm animals.

V: Do you accept my numbers?

M: Yeah, but I think you've oversimplified. Most humans have long lives compared to farm animals. How long does a farm animal live?

V: Most of the animals people eat are chickens, which live around 40 days before being slaughtered. Pigs and cows live much longer, but there aren't so many.[3] So I'd guess the average lifespan of a food animal might be around 45 days.

M: So only 45 days of suffering. We live much longer, so an average human life probably contains way more suffering. I bet that human history contains much more total suffering than the suffering caused by two years of factory farming.

V: Fair point. What about, say, the last 20 years of factory farming?

M: Still not sure. Remember that I think animal suffering might be less conscious than human suffering.

V: Well, we don't need to work this out exactly. A practice need not cause more suffering than all the suffering in human history, in order to be worth our while to change. Clearly factory farming has caused an *enormous amount* of suffering.

M: True.

V: And it's going to continue year after year if no one does anything about it.

M: Also true. I was just reacting to the idea that it's the world's greatest problem.

V: And I was reacting to your suggestion that we shouldn't worry about this until we've solved all

3 On farm animal lifespans, see Four Paws, "Farm Animal Life Expectancy," http://www.four-paws.us/campaigns/farm-animals-/farm-animal-life-expectancy/, accessed December 5, 2017. On the relative numbers of each type of animal, see One Green Planet, "Animals Killed for Food in the U.S. Increases in 2010," October 21, 2011, http://www.onegreenplanet.org/news/animals-killed-for-food-in-the-u-s-increases-in-2010/, accessed December 5, 2017.

human problems. It's open to debate whether factory farming has caused more suffering than all the suffering in human history, but either way, it's at least a reasonable candidate for being the world's worst problem. And even if it isn't the number one worst problem, it is still clearly incredibly horrible.

M: I'm sorry, but this is making you sound like a crazy extremist.

(f) On rejecting positions that "sound crazy"

V: Is there something wrong with my reasoning?

M: I don't know what's wrong with it, but the idea that animal agriculture is worse than the problem of war, or poverty, or disease, just *sounds* to me so extreme that it makes me want to say there must be something wrong with your argument.

V: And you think that's enough to reject the argument?

M: I do. I learned that from G. E. Moore: if you have an argument for a conclusion that seems crazy, you should reject it, even if you can't say exactly what's wrong with it.[4]

V: G. E. Moore was responding to philosophical skeptics who argue that no one knows anything about the world outside their own minds.

M: Right. The idea that I don't know, for instance, whether I have hands is so implausible on its face that if I hear an argument for that conclusion, I should infer that *some* step in the argument is wrong, even if I can't say which one or why.

V: And you think I'm like the philosophical skeptic.

M: Well, some of your views also sound crazy at first.

4 See, for example, G. E. Moore, "Hume's Theory Examined," pp. 108–26 in *Some Main Problems of Philosophy*, edited by H. D. Lewis (New York, NY: Macmillan, 1953).

V: Okay, let's examine that reaction. Would you agree that *sometimes* we should accept conclusions that initially sound crazy?

M: I don't know. What do you have in mind?

V: Here's an example I once heard. Imagine that you have a very large but very thin sheet of paper, one thousandth of an inch thick. You fold it in half, making it two thousandths of an inch thick. Then fold it in half again, making it four thousandths of an inch thick. And so on. After folding it fifty times, how thick would it be?

M: I don't know. Let me get out my calculator. (*pulls calculator out of backpack*)

V: First just take an intuitive guess.

M: Um . . . ten feet?

V: Sounds reasonable. Most people will agree that the answer is something under a hundred feet. What would you say if I told you that the correct answer is over 17 *million miles*?

M: That's crazy!

V: Yeah, it's crazy. But it's definitely correct. Enter it on your calculator. 0.001 inches, multiplied by two to the fiftieth power.

M: (*types on calculator, reads result*) 1.12×10^{12}.

V: That's the number of inches. To convert to feet, divide by 12. Then to convert to miles, divide by 5,280.

M: You're right, it's almost 18 million. But that's crazy.

V: Do you think your calculator is lying to you?

M: Of course not.

V: Do you think there must be something wrong with the argument because the conclusion is so crazy?

M: (*sigh*) No, it's correct. I'm not unreasonable, you know. It's just very surprising.

V: So sometimes we should accept conclusions that sound crazy.

M: Yeah, but that's a math problem. Ethical judgments are different.

(g) How ethics differs from mathematics and science

V: Different how?

M: I'm not sure. Maybe because ethical premises are less certain and less reliable than descriptive, factual premises.

V: Wouldn't that mean that your sense of what is "crazy" in ethics is also less reliable?

M: I guess so. But my point is that your ethical argument is less reliable than a mathematical calculation or a scientific claim or an observation of the physical world. That's why the sense of "craziness" could be enough to defeat an ethical argument, even though it wouldn't defeat a mathematical calculation, scientific theory, or physical observation.

V: Perhaps. But before we conclude that, let's first try to figure out where the craziness comes from.

M: What do you mean by "where it comes from"?

V: Sometimes, we can identify the particular point in an argument where things become surprising. Take the example of the folded paper. First I say that the thickness of the paper after fifty folds is 0.001 inches times 2^{50}. That statement isn't weird or surprising or controversial. What's surprising is just how enormous two to the fiftieth power turns out to be. *That's* where the "craziness" of the final answer comes from.

(h) Where does the "craziness" of extreme animal welfare positions come from?

M: Yeah, I was pretty surprised by that. I guess I'm not so good with large numbers. But I trust the calculator.

V: That's why it's not reasonable to conclude that there must be something wrong with the argument.

M: Okay, but how does this apply to your argument? You say factory farming is the worst problem in the world, which sounds crazy. Where does the seeming craziness of *that* conclusion come from?

V: Let's review the major premises in my reasoning. Some of them are *moral* premises, and some are *empirical, factual* premises.[5] First, I have the moral premise that suffering is bad. Anything surprising there?

M: No, that seems obvious enough. But I think it's surprising that animal suffering matters just as much as human suffering.

V: But I don't need to assume that. I can just say that animal suffering is at least one one thousandth as bad as qualitatively similar human suffering. That's enough for my argument. Is that surprising?

M: I'm surprised that an animal welfare nut would admit that humans might matter a thousand times more than animals.

V: Right, *that's* surprising. But it's not surprising that animal welfare matters at least one thousandth as much as human welfare, is it?

M: Not particularly.

V: The next step in my argument is just a factual, empirical premise: that life on factory farms is extremely unpleasant. Is that surprising?

M: You know, after we talked last time, I watched one of those PETA videos, "Meet Your Meat," about conditions on factory farms.[6] I had no idea how cruel they were.

V: So that part *is* surprising.

M: I guess so.

5 Empirical premises are ones that are based, directly or indirectly, on observation.

6 People for the Ethical Treatment of Animals, "Meet your Meat" (video), May 1, 2002, https://www.peta.org/videos/meet-your-meat/.

V: Here's my other factual premise: the number of animals killed in two years of factory farming is greater than the total number of humans who have ever existed. Were you expecting that?

M: Okay, that's surprising. I was kind of shocked to hear that.

V: That's where the "craziness" comes from. My moral claims aren't surprising; it's the empirical facts that are surprising. It's shocking that factory farming might be the world's worst problem, not because it's shocking that animal suffering might matter, but because the quantity of animal suffering is shockingly large.

(i) Questioning the vegetarian's empirical premises

M: So if I find your conclusion incredible, I should question the empirical claims about the quantity of suffering.

V: Do you think that would be a rational reaction to the argument?

M: Why wouldn't it be?

V: You can't use a moral assessment of some case to figure out what the empirical facts of the case are. That's because a moral assessment isn't reliable unless it is based on independently known empirical facts to begin with. For instance, your moral assessment of meat-eating isn't reliable unless it takes into account the empirical facts about the effects of meat-eating. Therefore, you can't figure out what those empirical facts are based on your initial sense that meat-eating isn't extremely wrong.

(j) Questioning the vegetarian's moral premises

M: I see. But then maybe I should deny one of your moral premises.

V: I don't think that would make sense either. That's why I made the point about how the surprisingness of my conclusions is due to the empirical facts, not my moral premises. You shouldn't reject an obvious moral principle based on your assessment of a particular case, if you didn't know the facts when you made that assessment.

M: That's a bit abstract for me.

V: Here's an illustration. Let's say you're initially extremely confident that Alice is a good person. You also believe that a good person wouldn't murder someone for money. Now suppose you learn, to your great surprise, that Alice has murdered someone for money. What should you conclude: (a) that Alice isn't a good person after all, or (b) that murdering for money is consistent with being good?

M: Obviously (a).

V: Good. That's like our case. You're initially convinced that meat-eating is okay, or at least not awful. You also believe that causing enormous suffering for trivial reasons is awful. Then you learn that meat-eating causes enormous suffering for trivial reasons. What should you conclude: (a) that meat-eating is awful, or (b) that causing enormous suffering for trivial reasons isn't awful?

(k) Biases in favor of meat-eating

M: I see your analogy. But why does your conclusion still sound hard to believe to me?

V: I can think of several plausible explanations for that.

M: Start with the biggest one.

(l) Status quo bias

V: Okay. Number one: status quo bias.

M: What's that?

V: It's a bias in favor of the practices that you're currently used to. If the people around you are acting in a certain way, and not getting criticized for it, then you tend to assume that behavior is normal and acceptable when you're forming your moral beliefs.

M: Oh, but I certainly don't think that everything that's commonly accepted is okay.

V: Few people do. The status quo bias isn't an explicit belief. The way it works is that the practices that we've gotten used to simply "seem normal" to us, and that colors the rest of our thinking. Often, we're unaware that this is happening. This is why people from different cultures wind up thinking radically different practices are right. And why a proposal to radically alter the practices may strike us as "crazy."

M: Well, maybe this tendency is a good thing. It's how we preserve our culture and traditions.

V: Perhaps it's a good thing in most cases. It stops you from stealing, driving on the wrong side of the street, and so on. But it can also lead to mass atrocities. Before the twentieth century, it led many people to accept slavery, to treat slave-masters with respect and runaway slaves as criminals.

M: So you think factory farming is the slavery of today.

V: I do. Our society has always had flaws and moral blind spots, which later generations look back at and shake their heads at. Slavery was one of them. It would be surprising, wouldn't it, if today was the first time in history when there weren't any major moral flaws in our way of life?

M: Sure, I guess. But that doesn't mean that factory farming is one of those flaws.

V: That's what we're trying to figure out. But on the face of it, the movement for animal welfare seems to fit the pattern of past moral progress.

(m) Speciesism

Much of the progress of the past was about overcoming prejudices against non-dominant groups – prejudice based on race, sex, religion, sexual orientation . . .[7]

M: And then prejudice based on species?

V: That's the next one.

M: But that's different. Sex and race differences are obviously morally irrelevant – your sex and race don't affect what rights you have, how important your interests are, or anything like that. But species differences are obviously relevant.[8]

V: Two centuries ago, people would have said that sex and race differences are obviously morally relevant. Back then, the abolitionists and the advocates for women's suffrage were the "crazy extremists."

M: When I introspect, it doesn't *seem* to me that I'm just accepting meat-eating because other people are doing it.

V: People are often mistaken about why they have the intuitions and beliefs that they do. We're unaware of many of the things that influence how we see things. But status quo bias is an extremely widespread and well-known phenomenon, so you are probably subject to it too.

M: But there are other cases in which I criticize the status quo. For instance, I oppose the current President and many of his policies.

V: True, but meat-eating is something you actually see done in front of you on a daily basis, by almost everyone. Government policies are just things that you hear about in the media.

7 See Peter Singer, *The Expanding Circle* (Princeton, NJ: Princeton University Press, 2011).

8 M again follows Richard Posner's views from his debate with Singer, *op. cit.*

M: So the status quo bias mostly applies to actions that you observe in your day-to-day life?

V: I think so. Also, your political views have more "social proof."[9]

(n) Social proof

M: What do you mean, "social proof"?

V: It's similar to status quo bias. People have a bias toward believing what other people believe – or at least things that are *close* to what *a good number of their peers* believe. A view that is too far out of the mainstream of opinion will tend to strike us as "crazy."

M: Well, you're definitely out of the mainstream. I note that this explanation only works if there are already other reasons why your view is unpopular; then this factor might contribute to making it even less popular.

V: Right. Another influence is self-interest. People tend to be biased in favor of beliefs that serve their own interests. For example, people in the slavery era who owned slaves had an interest in believing that slavery was okay. Otherwise, they'd have to give them up, at great financial cost. Plus, they'd have to believe unflattering things about themselves.

(o) Self-interest bias

M: Well, of course it's in my interests to keep eating meat, and I'd prefer to think that it was okay while I'm doing it. But it doesn't seem to me that I'm being influenced by that.

V: It probably wouldn't. Most people who are influenced by a bias can't themselves detect the bias. You have this sense that my conclusion is "crazy," and you don't know why it seems

9 See Robert B. Cialdini, *Influence: The Psychology of Persuasion* (New York, NY: William Morrow & Co., 1993), ch. 4.

that way. So the seeming is probably caused by some unconscious factors like these.

M: I see. Is that all?

(p) Empathy and the affect heuristic

V: Not yet. Another factor is something called the "affect heuristic."[10] It's the tendency to evaluate how good or bad something is by reference to the strength of the emotional reaction we feel when we contemplate the thing.

M: That doesn't sound so wrong. Usually, the worse something is, the worse I feel about it; the better it is, the better I feel.

V: Usually, yes. But there are at least two reasons why we might go astray in the case of animal ethics. One is that our capacity for empathy with other species is limited. We find it harder to empathize with other species than with other humans. So we have diminished affective reactions when we think about animal suffering, compared to human suffering.

M: Okay, but maybe the explanation goes the other way: maybe I have diminished empathic responses to animal suffering *because* I know that animal suffering is less important than human suffering.

V: I think that's unlikely; I have a better explanation: human beings evolved as social animals. The capacity for empathy probably evolved to

10 "Heuristics" are mental shortcuts for coming up with answers, which usually work reasonably well but can sometimes lead us astray. "Affect" in psychology refers to feelings. For discussion of the affect heuristic as used in decision-making, see Paul Slovic, Melissa Finucane, Ellen Peters, and Donald G. MacGregor, "The Affect Heuristic" in *Heuristics and Biases: The Psychology of Intuitive Judgment*, edited by Thomas Gilovich, Dale Griffin, and Daniel Kahneman (Cambridge, UK: Cambridge University Press, 2002), pp. 397–420. On the application of the affect heuristic to moral philosophy, see Michael Huemer, "Transitivity, Comparative Value, and the Methods of Ethics," *Ethics* 123 (2013): 318–45, at pp. 328–30.

enable us to get along better with other humans in our social group, not to get along with other species. That's why we don't empathize as readily with other species.

M: That sounds speculative.

V: True. I'm just trying to offer plausible explanations for your intuitive reactions.

M: Okay. What's the other problem with the affect heuristic?

(q) Problems with intuitions concerning large numbers

V: We have a well-known problem dealing with large quantities. We can't intuitively grasp them. Moreover, as we imagine larger quantities of something that's good or bad, our affective response doesn't increase proportionately.

M: I suppose that's a good thing. Otherwise, we'd be in constant emotional agony from following the news.

V: Right. If you hear about a disaster that killed five people, you feel sorry about that. If you hear about one that killed five *thousand* people, you don't feel a thousand times as sorry.

M: Do we just feel the maximum degree of sorriness then?

V: Not even that. If there is a maximum intensity of negative emotion, we don't necessarily feel it even in response to colossal evils. A vivid description of one death, by a sympathetic person, might make us feel worse than a report of a million deaths. Many factors affect our emotional response. It clearly isn't simply proportional to the size of the good or bad event – not even close.

M: So this leads us astray when we try to assess the badness of large evils.

V: Yeah, like when we talk about something happening to *billions* of creatures. Our minds can't really appreciate, or respond proportionately, to such quantities. Harming a billion creatures is a thousand times worse than harming a million creatures – but we don't *feel* that way.

M: But this doesn't seem to lead us astray much when we are thinking about human harms. If I ask someone, "How much worse is it to kill a billion people than to kill one person?", I bet most people would get the correct answer: one billion times worse.

V: Probably. But that's because they don't have to rely on an independent moral intuition to make the comparison – they just look at the numbers. It's when you compare bads belonging to different categories that you deploy the affect heuristic.

M: Different categories? So like, if I'm asked to compare a broken promise to a sprained ankle?

V: Right. Or animal suffering to human suffering.

M: But according to you animal advocates, those *are* in the same category.

V: But *most* people *think* of them as belonging to different categories. Most people think you have to deploy an independent moral intuition to compare animal and human harms. So they do – and that intuition is affect-driven.

M: So you're against relying on affect in moral evaluation?

V: Not necessarily, not across the board. It's just that in some cases we can predict that it would be unreliable.

M: But in order to say that the affect heuristic is unreliable in this case, don't you have to already know what the correct moral judgment is? If it's leading us away from the truth, it's unreliable; but if it's giving us the correct answer, then it's reliable.

V: No, when I say it's unreliable, what I mean is this: we shouldn't have any *independent expectation* that it would get us the truth. The affect heuristic would lead us to judge human interests more important than animal interests, whether or not that was true. So, if you start out with no opinion about whether human interests matter more, you can't trust your emotional responses to tell you the answer.

M: Okay, so couldn't the large-numbers problem be avoided by just asking people to think about individual cases? Like, imagine a single cow suffering on a factory farm for a day, then imagine a single person enjoying the pleasure of a hamburger. We should be able to compare those two, right?

V: You're right, that would avoid the large-numbers problem, though it's still subject to the other sources of bias I mentioned.

M: Well, when I think about the cow on the factory farm, it doesn't seem very bad to me.

V: Really? When I think about it, it seems very bad to me – clearly much worse than someone being deprived of the pleasure of a hamburger.

M: I guess we have a basic clash of intuitions. I wonder why we have such different reactions.

V: People vary in their capacity for empathizing with other species.

M: That's true, I find it pretty hard to empathize with a cow. But why should I trust your intuitions, rather than my own?

V: Remember all the biases we were just talking about? Status quo bias, social proof, self-interest, the difficulty of empathizing with other species . . . ?

M: Sure. But your intuitions are also biased.

V: How do you figure that?

M: You just admitted it: you empathize with non-human animals. That's biasing your moral judgment.[11]

V: I said that was explaining the difference in our reactions. I didn't say it was a bias on my part.

M: You don't think empathy can function as a bias?

(r) Empathy and psychopathy

V: I don't see any reason to think it's a bias in this case. Compare another case: the case of psychopaths. Psychopaths lack the capacity for empathy in general. Does that mean that they make the most objective, unbiased moral judgments?

M: I'm not sure they make moral judgments at all.[12]

V: Right, their lack of empathy prevents them from taking others' experiences into account. It doesn't make them objective; it makes them ethically blind.[13]

M: Okay, obviously a complete lack of empathy is a problem. But too much empathy can also be a problem. I know someone who has too much empathy, and it messes up her life. She feels a lot of anguish because of other people's problems. She's even gone into serious debt trying to help others.

V: Yeah, that sounds like a practical problem. But I'm not sure it's relevant to the point here.

M: Why not? You were talking about how important empathy is.

11 On the biasing effects of empathy, see Paul Bloom, *Against Empathy: The Case for Rational Compassion* (New York, NY: Ecco, 2016).

12 For discussion, see Walter Sinnott-Armstrong, "Do Psychopaths Refute Internalism?" in *Being Amoral: Psychopathy and Moral Incapacity*, edited by Thomas Schramme (Cambridge, MA: MIT Press, 2014), pp. 187–207.

13 Iskra Fileva ("Reflection Without Empathy," unpublished manuscript) argues that psychopaths are unable to reason morally due to their incapacity for affective empathy.

V: Yeah, but I'm not saying empathy is good in all ways and in all contexts. Of course, it's not necessarily good *from the standpoint of self-interest.* What I'm saying is that empathy helps us to perceive morally relevant factors that depend on the interests of others; without it, we just care about our own interests. So your overly empathic friend is doing poorly with respect to promoting her own interests, but she's probably doing quite well with respect to appreciating the moral relevance of others' interests.

M: Point taken. But too much empathy can also lead to *moral* errors. For instance, we might give money to charities that help people in a visible way – like those ones where you sponsor a child and they send you pictures of the child and stuff – instead of giving to charities that don't send you pictures but that are actually more cost-effective. That's because of empathy.

V: Yeah, that's true too. So empathy isn't *sufficient* for making good moral choices, but it might be *necessary.* You need it in order to be moved to take account of interests other than your own. You still need to use reason to decide what to do about those other interests, but if you lack the capacity for empathy, you'll just ignore others' interests. Like how psychopaths just ignore other people's interests, and humans just ignore the interests of animals.

M: But I don't lack the capacity for empathy; I just have more trouble empathizing with other species than with my own – which is perfectly normal.

V: That's normal, true. But it's also true that it prevents you from fully taking other creatures' experiences into account. That doesn't make you more objective; it makes you less aware.

(s) Avoiding dogmatism

M: Well, you've made some interesting points that I'll have to think about, V. But I'm not yet convinced. Maybe your arguments are pieces of sophistry that I'm just not clever enough to see through.

V: The G. E. Moore shift again? I thought we already discussed why that isn't a rational response.

M: I know, but maybe the arguments you gave to show why that isn't a rational response were *themselves* just pieces of clever sophistry.

V: If you're going to say stuff like that, there's no way I could ever convince you. Whatever I say, you can always say that maybe I'm wrong for some reason you can't identify. That's called being dogmatic.

M: No, I'm not being dogmatic. I'm not saying you're definitely wrong. I'm just saying I'm not fully convinced. And I'm not saying I'll never be convinced; I just need to think about it more.

V: Okay, so for now, you're not sure whether it's wrong to eat meat. Do you think it might be *obligatory* to eat meat?

M: Don't be silly. I just mean that I think it might be morally okay.

(t) Erring on the side of caution

V: Alright, it might be wrong, or it might be just okay. In that case, I would suggest that, until you figure out which it is, maybe you should stop doing it. If there's even a fair chance that it's extremely wrong, better stop until you're more sure. You want to be on the safe side, right?

M: In general, yeah. But I can't avoid every action that *might* be wrong. I mean there's *some* chance that just about anything I do might be wrong. But I can't be avoiding everything.

V: Fair enough. But I'm not asking you to avoid every action that merely has some non-zero probability of being wrong. I'm saying: avoid an action if it has a *pretty good* chance of being *very* wrong, where you have no moral reasons *to* do it, and where you can avoid it without unreasonable personal cost.

M: Well, that's hard to object to. But until I've finished thinking through all the arguments, I'm not sure if I should even say there's a "pretty good chance" that you're right.

V: I think you know enough to say there's at least a pretty good chance. You know that the issue turns on a moral intuition about the badness of animal suffering. This intuition is held by many people who appear to be in general reasonable, smart, and morally sensitive.

M: I guess that's fair to say.

V: In fact, many of them consider the intuition extremely obvious. The great majority of the literature in ethics on the topic also agrees that meat-eating in our society is generally wrong. Many of these experts consider the case decisive.[14]

M: But most people in our society seem to think eating meat is fine. And even most philosophers seem to be okay with it.

V: Right, so there's a divergence between ethicists who work on the topic, and laypeople or philosophers who work in other areas. Now all of this that I just said – this is all stuff that you can know, independently of your direct evaluation of the arguments. I mean, you don't have to first figure out what you think of the arguments, to

14 See Rachels, "Vegetarianism," *op. cit.*, pp. 884, 898.

know that most ethicists who work on the topic think meat-eating is wrong.

M: Okay, but we shouldn't just defer to the experts on a controversial topic like this.

V: Yeah, I'm not saying we should just defer to the experts. I'm saying that the opinion of these experts, together with the plausibility of the arguments we've been discussing, is enough for you to say that there's at least a pretty good chance that I'm right – at least until you can come up with a good argument against my views.

M: Maybe you're right. But going vegetarian is going to make my life so much worse. I can't commit to such a big lifestyle change.

V: How about you try being vegetarian just for the next week, and then we'll talk about it here at the same time next week?

M: (*sigh*) Oh, alright. I hope you appreciate the big sacrifice I'm making for you, V.

Day 4 The Vegan Life, Abstract Theory, and Moral Motivation

Setting: Same place.

V: Hi, M. How have you been? Have you tried out the vegetarian life?

M: I tried it for the last week, like you insisted.

V: Well, how did it go?

M: Oh, it was miserable! You know how much I like food. Life is barely worth living without fried chicken!

V: Sorry, it's partly my fault. I could have left you with some advice on how to find yummy plant-based meals.

(a) Finding good vegan meals

M: *Are* there any such things? I can't keep eating piles of broccoli and lettuce.

V: Of course there are. You seem to like the meals *here* well enough.

M: Okay, that's true, this restaurant is pretty good.

V: Most vegetarian restaurants are. They seem to put more thought into their food than the average conventional restaurant.

M: Yeah, but I can't afford to keep eating out.

V: That's understandable. Fortunately, you can eat at home for a lot less if you go vegetarian, since vegetables are cheaper than meat. You know, I have a great carrot ginger soup that I make at home. I'll send you the recipe.

M: Okay, but I can't be relying on you for advice every time I want to cook a meal. Plus, I'm busy; I can't spend a whole lot of time on cooking.

V: Just type "easy vegan meals" into a search engine any time you want a new idea. You'll get plenty of great suggestions.

(b) Avoiding eggs and dairy

M: Wait, wait. Did you say *vegan*? I thought I just had to give up meat. You mean I have to give up eggs and dairy too?

V: They come from factory farms too. If you buy them, you're paying people to do factory farming.

M: But I love cheese!

V: You can get vegan cheese.

M: Vegan cheese? What the heck is that?

V: It's a product that looks and tastes like cheese, but with no animal products. Usually made from cashews.

M: I'm skeptical.

V: You remember the cheese we had at the reception on Friday after the philosophy colloquium?

M: I remember. That stuff was good. You see, that's why I can't give up dairy.

V: That was vegan cheese.

M: Really? Are they all that good?

V: Some are better than others. Just find the ones you like. There are vegan substitutes for most animal products.

(c) Eating bivalves

M: But where am I going to get enough protein in my diet?

V: Beans, nuts, peanut butter. Also, bivalves.

M: What's a bivalve?

V: You know, clams, mussels, scallops.

M: Wait a minute. You eat clams? Clams are animals! You're no vegan. You're not even a vegetarian!

V: Bivalves have no brains.

M: So what? They're still alive, and people have to murder them to satisfy *your* appetite. You're no better than me with my chicken and hamburgers!

(d) The value of life

V: I don't think you understood any of the preceding arguments. None of the arguments were about life. After all, plants are alive too. So are bacteria; so are cancer cells.

M: So you don't think that life has intrinsic value?

V: I'm not making any claims about that. My argument was simply that it's wrong to inflict severe pain and suffering for no good reason.

M: So why do you think it's okay to inflict pain on plants and clams, you evil oppressor?

(e) Why can we eat plants?

V: Um, I don't think that. I think it's impossible to inflict pain on plants or clams.

M: How do you know that?

V: Pain is caused by electrical impulses from nerve endings reaching your brain. If the impulses don't reach your brain, you don't feel any pain. That's known empirically. So if you didn't even have a brain, then you couldn't feel pain at all.

M: Sure, that may be the current scientific understanding. But you can't be 100% certain of that. Maybe there's some other way to have pain, and plants and clams are having pain all the time.

V: And maybe the chair you're sitting on is in great agony. No way to prove it isn't. But we have no reason to think so, and we have to sit somewhere.

M: Okay, I don't really think plants are conscious. I was kidding with you. But I don't see why someone couldn't make an argument similar to your earlier argument from risk, to show that we can't eat plants.

V: How would that argument go?

M: You know, we're killing so many plants every year that if there's even a 1% chance that plants feel pain when they're killed, we have to stop the practice.

V: And the person making this argument would be saying that we should not eat anything?

M: Maybe – maybe we're ethically obligated to commit suicide.

V: Okay. Then here's how that argument would be different from my argument. Number one, that argument is asking us to sacrifice our lives. My argument is only asking you to give up a little pleasure at meal times. And that's because I think it's wrong to inflict suffering *for no good reason.*

M: So it's okay to hurt others in order to survive, but not to get more pleasure for yourself?

V: In general, it's much more likely to be okay. Now, here's the second difference. It is virtually certain that animals feel pain. That's clearly over 99% probable. But it is also virtually certain that plants *don't.* Since plants have no nervous systems, the probability that they feel pain is very much lower than 1%.

M: Okay, but you've been misusing that word "vegetarian." "Vegetarian" means a person who eats only plants. Clams and oysters aren't plants.

V: I don't care what word you use. You can call it schmegetarianism, compassionate eating, or whatever you want.[1] As long as you stop supporting animal cruelty.

1 The technical term is "ostroveganism."

(f) Killing insects

M: Alright, but what about insects? Plant farming kills trillions of insects every year, mostly with pesticides. Surely we're going to have to stop killing bugs, right?

V: Shouldn't I make the same points here? One, the costs of giving up killing insects are much higher than the costs of giving up meat-eating. Two, it's much less likely that insects feel pain. Plus, number three, we aren't raising insects in horrible conditions for their whole lives before killing them, as in factory farming. And four, animal farming requires killing insects, as well as plants, *in addition* to the animals directly killed for meat. So plant farming is still better.

M: Wait, how does animal farming kill insects and plants?

V: To raise an animal, you have to feed it vegetable foods, which come from farms. The amount of food you then get out of the animal is less than the amount of food that went into raising it. So any problem with plant farming is also a problem with animal farming – in addition to all the problems with animal farms that we've already discussed.

M: What do you mean about the costs of giving up killing insects being much higher?

V: Virtually all of modern life kills insects. You can't drive a car without killing some; you can barely walk without killing them.

M: Okay, so maybe giving up all killing would be too demanding. But surely we should still give up, say, using pesticides on farms, right? Because we could buy organic foods instead.

V: If you want to argue for that, go ahead. But I think you're changing the subject. Let's first agree to give up buying factory-farmed meat. Then we can worry about more controversial cases.

(g) Are vegetarians hypocritical?

M: But I've seen you eating conventionally raised vegetables! Right there! (*points at V's lunch*) I bet some of those vegetables are conventional!

V: Now you're resorting to argument ad hominem. You're accusing me of being bad or hypocritical or something, instead of just focusing on the arguments.

M: Shouldn't I expect the proponents of a theory to be consistent?

V: Whether or not I personally act in the morally best way is irrelevant to the truth of the moral principles we've been discussing. Even if I'm hypocritical as you claim, that wouldn't make it okay for you to keep buying meat. It's just a way to distract yourself from the moral issue about your own behavior.

(h) Unconscious speciesism

M: But I don't think it's irrelevant, because I think your failure to care about insects shows that you don't really believe the moral views you've been advancing.[2]

V: If I don't believe them, then why have I been eating vegan for the last three years? And why have I been telling you all these arguments? Am I just a crazy liar?

M: Okay, you believe them on one level. But on another level, you don't *completely* believe them, because you haven't fully integrated them into your thinking and your lifestyle. If you really, fully believed that species membership doesn't matter, then you'd care more about insects.

2 Cf. Bryan Caplan, "Bugs," Econlog, October 3, 2016, http://econlog.econlib.org/archives/2016/10/bugs.html.

V: So you think I'm an unconscious speciesist?

M: That's a good way to put it.

V: Maybe. There are also experiments that suggest that most of us are unconscious racists.[3] But that doesn't show that such attitudes are correct or morally okay.

M: No, it doesn't show that. But one explanation for why you haven't rooted out all your speciesist prejudices is that they're actually correct.

V: I guess that's a possible explanation. But you'd have to think that unconscious biases are more reliable than conclusions of conscious reasoning. And we talked last time about the biases that may be distorting our judgment.

M: Alright, maybe your behavior doesn't give us a good reason for rejecting your philosophical arguments. But I still want to know: *should* we be a lot more careful about protecting insects?

V: If insects were sentient like cows, then I'd say sure.

(i) Are insects sentient?

M: Why don't you think insects are sentient? They've got eyes and other sense organs, so they must have sensations.

V: Three reasons. One, they don't have nociceptors –

M: What? "Noss receptors"?

V: Nociceptors. The kind of nerve cells that sense pain. They don't have 'em. Second, they have drastically simpler central nervous systems. Like a hundred thousand times simpler.

M: Maybe you only need a simple nervous system to have pain.

3 Here, V alludes to the literature on implicit bias; see Anthony G. Greenwald and Mahzarin R. Banaji, "Implicit Social Cognition: Attitudes, Self-Esteem, and Stereotypes," *Psychological Review* 102 (1995): 4–27.

V: But you're going to have a hard time explaining the third point: insects don't show normal pain behavior. An insect with a crushed leg keeps applying the same force to that leg. Insects will keep eating, mating, or whatever they're doing, even when badly injured – even while another creature is eating *them*.[4]

M: Okay, so there's a pretty good case that they don't feel pain. But there's still *some chance* of it, right?

V: I think you're engaging in distraction again. If you want to become a bug activist, go ahead. But first, let's agree to give up the much more clearly wrong practice of buying meat from factory farms.

(j) Free-range and humane certified meat

M: Alright, maybe I should give up factory farm meat. What if I just buy free-range meat instead?

V: Unfortunately, companies can legally call their products "free-range" and still have a lot of cruelty. If you want humanely made products, you have to look for things that are certified by an animal welfare organization. For example, you can look for the "Certified Humane" logo.[5]

4 See C. H. Eisemann, W. K. Jorgensen, D. J. Merritt, M. J. Rice, B. W. Cribb, P. D. Webb, and M. P. Zalucki, "Do Insects Feel Pain? – A Biological View," *Experientia* 40 (1984): 164–7.

5 See http://certifiedhumane.org/.

M: Then it'll be ethical to buy it?

V: Then it *might* be ethical. Buying from factory farms is clearly wrong; buying humane certified meat might be acceptable. Depends on whether you believe in animal *rights* or not.[6]

(k) Animal rights

M: Well, how come we haven't been talking about that?

V: I focus on factory farming because it's the source of nearly all animal products. I figure I should first try to dissuade people from doing the clearly terrible thing that almost everyone is doing every day (buying from factory farms), before worrying about something that a tiny minority of people are doing that's much less bad but *might* still be unethical (buying from humane farms).

M: I see. But do *you* buy humane certified meat?

V: I don't buy it because I don't know if it is ethical. I figure that if I don't know, I shouldn't do it.[7]

M: Why don't you know?

V: Well, I'd have to figure out whether it's permissible to kill animals humanely for food. For that, I'd have to figure out whether they have a right to life. And for that, I guess I'd have to first figure out what's the basis for the right to life in general.

M: Isn't that what we have moral philosophers for?

V: Yeah, but the moral philosophers don't agree.

6 For a defense of animal rights, see Tom Regan, "The Moral Basis of Vegetarianism," *Canadian Journal of Philosophy* 5 (1975): 181–214, available at http://tomregan.free.fr/Tom-Regan-The-Moral-Basis-of-Vegetarianism-1975.pdf. For an argument that humane animal farming is wrong even if animals lack rights, see Jeff McMahan, "Eating Animals the Nice Way," *Daedalus* 137(1) (Winter 2008): 66–76.

7 Here, V follows the reasoning of Rachels, "Vegetarianism," *op. cit.*, p. 894.

M: Professor Tooley told me that the right to life is based on one's conception of oneself as a subject of experience continuing through time.[8]

V: That's one theory. Another view is that the right to life rests on one's being the subject of a life that matters to oneself. Or perhaps it rests on one's having the potential for a human-like future. Or perhaps there aren't any such things as rights in the first place.[9]

(l) Debating the correct ethical theory

M: Why don't we just figure out which theory is true?

V: Easier said than done. The leading experts can't agree, so it seems unlikely that we can settle it here. If we start on that, we'll just argue about that forever.

M: Don't be so pessimistic. Maybe we're better at this than the leading experts.

V: Really? Remember that time when we started talking about the definition of knowledge?

M: Yeah, that was a great conversation.

V: A great conversation that went on for three hours. It only stopped because you had to leave.[10]

M: Yeah. I still want to raise more objections to your last theory on that . . .

V: And then there was the time we started talking about free will.

M: Another great conversation.

8 See Michael Tooley, "Abortion and Infanticide," *Philosophy and Public Affairs* 2 (1972): 37–65.

9 For these views, see, respectively, Tom Regan, "The Moral Basis of Vegetarianism," *op. cit.*, and *The Case for Animal Rights*, *op. cit.*; Don Marquis, "Why Abortion Is Immoral," *Journal of Philosophy* 86 (1989): 183–202; and C. L. Sheng, "A Defense of Utilitarianism Against Rights-Theory," *Social Philosophy Today* 5 (1991): 269–99.

10 Philosophers have debated the definition of knowledge for many years, with no consensus. For a review, see Robert K. Shope, *The Analysis of Knowing: A Decade of Research* (Princeton, NJ: Princeton University Press, 1983).

V: That went on for five and a half hours, and we never reached any agreement.

M: Okay, so we're probably not going to deduce the correct ethical theory. But how can we know buying meat is wrong, without knowing the correct ethical theory?

V: Because the argument against meat rests on intuitive, very widely shared moral beliefs, like "it's wrong to inflict a lot of suffering for no good reason" and "it's wrong to pay people for immoral behavior." Any reasonable ethical theory is going to agree with those.

M: Okay, but all of your arguments assume that there are objective moral truths, don't they? *That's* controversial.

(m) Are there objective values?

V: Do *all* moral arguments assume that there are objective moral truths?

M: I don't think so.

V: Then mine doesn't either.

M: Why do you say that?

V: Because I don't see how my argument differs from any ordinary moral argument. It's up to you to tell me how I presupposed objective values in a way that other moral arguments don't.

M: Maybe because you're arguing for a radical revision of our practices. On some theories, morality is just based on social practices.

V: Could that view have been used to defend slavery, back when that was the practice? The oppression of women? Medieval torture?

M: I guess it might.

V: Okay, if my argument against meat-eating is only as strong as the arguments against slavery were – if the only people who should disagree with me are people who think we had no reason to give up slavery – I'm okay with that.

M: Well, maybe slavery was wrong, even though it fit with the practices of the time, because it conflicted with some deeper values held by society.[11]

V: Like what?

M: Maybe there were deeper values of liberty and autonomy, and a consistent application of those values, without making arbitrary distinctions, required granting freedom to slaves.

V: In that case, I think factory farming is also wrong because it conflicts with deeper values of our society. Like the deeper values of compassion and "not inflicting needless suffering." That's just as plausible as the story about slavery.

(n) Moral skepticism

M: I guess so. But so much seems uncertain in this area. I'm not sure it's worth changing my lifestyle, when the experts can't even agree on whether there are any moral facts, whether they're dependent on conventions, and so on.

V: This sort of skepticism only seems to come up when people are criticized for behavior that they don't want to change, and they run out of ways of trying to defend it. Only then do you start feeling skeptical about morality. The rest of the time, you have no trouble accepting moral judgments.

M: What do you mean? I don't go around judging people all the time!

(o) Why philosophers should not serve on juries

V: Let me give you an example. Say person A is suing person B, and we're on the jury. We're supposed to decide if B wronged A in a way that demands compensation.

M: Isn't the jury just supposed to decide whether B did something illegal?

11 Following Harman, *The Nature of Morality*, *op. cit.*, pp. 94–5.

V: Just assume that the law says A is entitled to compensation only if B wronged A.

M: Okay.

V: It turns out that what B did was to smash A's car with a sledgehammer, just for fun, causing $2000 worth of damage. Several witnesses saw it.

M: Sounds like an easy case. A gets $2000.

V: Not so fast! There are a few philosophers in the jury room: a metaphysician, a political theorist, an epistemologist, and an ethicist. The metaphysician argues that B isn't responsible for his action, because there's no such thing as free will.

M: I guess that could make sense . . .

V: The political theorist says that B's action wasn't wrong because property rights are illegitimate. The epistemologist says that we can't accept the eyewitnesses' testimony until we first prove that the senses are reliable. Finally, the ethicist says that there are no moral facts, so B can't have done anything *wrong*.

M: I guess this is why they don't usually allow philosophers on the jury.

V: (*laughs*) No doubt. So how would you vote?

M: If I agreed with one of those philosophers, I'd have to support the defendant.

V: Right. But how would *you actually* vote? Would you say B did nothing wrong?

M: No. Personally, I'd still vote to award $2000 to A.

V: So skeptical philosophical theories don't prevent you from making moral judgments about other people's behavior.

M: No.

V: In fact, when I first told you the story, you said it sounded like an easy case.

M: Yeah, it did.

V: Well, the case of ethical vegetarianism is just as easy. There's no more doubt about the wrongness of meat-eating than there is about the wrongness of smashing someone's car for the fun of it – or the wrongness of beating children, or killing people for money, or any other paradigm wrong. You wouldn't go ahead and do those other things just because there might not be any moral facts, would you?

M: No. But you really think being vegetarian is a simple, clear decision, just like deciding not to kill people for money?

V: Basically, yes. At its heart, the question is: do I support something that causes enormous pain and suffering, for the sake of minor benefits for me? That's it. It's not "Are human lives more valuable than animal lives?" It's not "Are there objective values?" or "Are there rights?" or "What's the basis for rights?" It's just about causing great suffering for small gains.

(p) Is giving up meat "too difficult"?

M: Well, your arguments sound reasonable and all. But I just don't think I can give up meat. It's too difficult, and I am weak-willed.

V: I don't think that's true.

M: I'm telling you, I'm not ready to give up meat. You think I'm lying?

V: I don't think you're lying. But people are often mistaken about why they do what they do. If you keep eating meat, it won't be because it was too difficult to give it up.

M: What do you mean?

V: Imagine the next time you're in a restaurant, and you're thinking of ordering a meat dish. Imagine I show up just before you order, and I offer you $20 to order a vegetarian dish instead.

M: Twenty bucks? That would be like getting a free meal at most places, plus some extra cash.

V: Would you take it?

M: Sure.

V: So it's not really very difficult to refrain from eating meat. It can't be very hard, if just paying you $20 gets you to do it.

M: Sure, it would be easy to forego one time. But after going for a few weeks without it, it would get harder, as I started to miss the taste. Don't you find it a terrible struggle to stay vegan?

V: No, not really.

M: You must have an iron will. Day in and day out, seeing delicious meat around you . . . What's your secret?

V: I don't deliberate. I don't decide, every time I eat, whether to be vegan. If I had to decide every time, I imagine that it would be hard.

M: Well, why don't you deliberate about it? I thought about it many times over the past week!

V: I don't deliberate about whether to do things that are wrong. I decided once, three years ago, that buying meat was wrong, and I've had no reason to reopen the issue. So I don't have to struggle with any decision.

M: Oh, I bet you do some wrong things from time to time. Come on.

V: Let me rephrase. I don't deliberate about doing things that I see as *terrible.* Stealing paper clips, sure. Mugging people? No way, not even thinking about it.

M: I don't know if it's possible for me to get in that state of mind.

V: When you see an attractive person, do you have to struggle to decide whether to grope them?

M: I would never do that.

V: Or when someone pisses you off, do you have to struggle with whether to smack their face?

M: Not that either!

V: Right. You already know that you don't do things like that, so there's no deliberation and no internal struggle.

M: I just haven't been able to get in that state of mind with this issue. This past week was so hard. I'm really hankering for a steak now.

V: Here's another hypothetical. You've gone without meat for three weeks, so you're hankering for a steak. Like now, only more so. We're out to lunch with Professor Carney, and he's got a steak on his plate. He likes to cut his food up into bite-sized pieces. You could reach over, spear a piece with your fork, and eat it before he can stop you. Assume that this is the only way for you to get some steak. Do you do it?

M: Hmm, would that be unethical, on your view? He's already ordered it . . .

V: Never mind that. I'm not asking what you *should* do. I'm asking a psychological question. Do you think you would in fact grab a piece of Carney's steak?

M: I think he would get mad if I did that.

V: Okay, what if he's gone for a bathroom break, so he won't even see you?

M: He might notice when he gets back.

V: He probably won't notice one piece. Even if he does, what's he going to do?

M: Well, maybe I *would* take one then!

V: Come on. Have you ever done something like that?

M: Okay, fine, I wouldn't take it.

V: Why not?

M: I don't know, people just don't do things like that.
V: Even though you have this supposedly irresistible need for steak?
M: Maybe you're just showing the power of social conventions. I have two powerful urges, but the urge to follow conventions is so strong that I can't take the steak.
V: Okay. Let's say you ask Bob for a piece of his steak. He offers to sell it to you for $20. Nothing wrong with accepting that offer, as far as society is concerned, right?
M: It's pretty weird, but not exactly socially unacceptable.
V: You happen to have a twenty in your pocket. Do you hand it over?
M: $20 for one bite? That's totally unreasonable.
V: What about your powerful, nearly irresistible steak urge?
M: One bite wouldn't satisfy it anyway.
V: Okay. What if he offers to sell you the whole steak for, say, $100?
M: Um . . . that's a bit much.
V: Notice how this is different from, say, a drug addict. The drug addict really would hand over $100 to get his fix. That's because addicts really have overpowering urges. You, however, do not.
M: Okay, fine. I guess it's not all *that* difficult to refrain from eating meat. I guess it's just that I'm a selfish jerk. Thanks for pointing that out.

(q) Are we too selfish to give up meat?

V: I don't think that's it either. Here's another hypothetical. You're in the library late at night. You see a desk where another student has been studying. The student has gone off to the bathroom and left his backpack there, with his wallet sitting on top. You could grab the wallet and leave, and no one would catch you. Do you take it?

M: Hey, I'm no thief!

V: Good. But a real selfish jerk would take it, right?

(r) Social conformity and the enforcement of morality

M: I guess so. Then what are you saying explains my choices?

V: Social conformity. Stealing is disapproved of in our society, and you've internalized that. Meat-eating isn't, and almost everyone around you is doing it. That's why you keep eating meat, no matter how wrong it is, but you won't steal, no matter how much it would benefit you.

M: So it's mostly peer pressure?

V: Yes. Now you understand why I was getting all "judgy" on you earlier.

M: You've been trying to exert peer pressure on me?

V: That's the main way morality is enforced in human societies. Other people express disapproval of bad behavior, and it makes us want to avoid that behavior. If there's no disapproval, then most of us keep doing it. That's why the people who know that a practice is wrong have to keep saying so.

M: You're making humans sound kind of weak. Like we just do whatever we see other people doing, however wrong it is.

V: Pretty much. Think about cases like Nazi Germany. Ordinary, average Germans, who never committed a crime before, found themselves helping to mass-murder Jews. The natural resistance to killing was overcome, mainly by social pressure.

M: Yeah, Germans back then sucked, didn't they?

V: Not just Germans back then. If you lived in the American South in the 1800's, you would probably have accepted slavery as perfectly natural.

M: Yeah, you talked about that before, when you were talking about the status quo bias. But what if you've already accepted that meat-eating is wrong, but you just can't give it up?

V: That happened with slavery too. Thomas Jefferson was a big philosophical opponent of slavery, yet he was also a slave-master.[12] If Jefferson were alive today, do you think he would still have slaves?

M: I assume not. Only a terrible person would hold slaves *today*. But what are you getting at?

V: It's about social conformity. Jefferson "couldn't give up" his slaves, not because he had some powerful urge to be a slave-master, and not even just because it would be so much against his interests (though it would have been), but because other people in his society had slaves and accepted the practice – that undermined his moral motivation. If he lived today, he wouldn't dream of owning other people, because it's so uniformly disapproved.

M: Is all this supposed to help me give up meat?

V: Yeah. If you think that you can't do it because you have these overpowering carnivorous urges, or even that you're just utterly selfish, then it's unlikely that you'll make the effort. But once you realize that you make comparable sacrifices to your interests all the time, and it's not that difficult, then you're more likely to do it. The reason you make other sacrifices but you're not making *this* sacrifice is a really bad reason: not enough other people are pressuring you.

12 For Jefferson's views on slavery, see Christa Dierksheide, "Thomas Jefferson and Slavery," The Thomas Jefferson Encyclopedia, 2008, https://www.monticello.org/site/plantation-and-slavery/thomas-jefferson-and-slavery.

(s) Are vegans too moralistic?

M: You know, you vegans are really preachy and moralistic. I'm not sure I want to be like you.

V: Well, I hope you understand now why we are so "moralistic," as you put it.

M: Because you have a character flaw that makes you take pleasure in controlling other people and feeling superior to them?

V: God, no. I *wish* I didn't have to do any of this. I wish the meat industry were ethical so I could buy their products. Or at least that everyone knew it was unethical so I wouldn't have to keep talking about it and having tensions with other people.

(t) How meat-eaters react to vegans

M: You mean being vegetarian causes other people to get mad at you?

V: Not exactly. Most people respect my dietary choices as long as I'm quiet about it. It's when I start talking about how *they* should stop eating meat that people get angry.

M: What do they do?

V: Sometimes they talk about how vegans are excessively moralistic or self-satisfied[13] – discounting the possibility that vegans sincerely care about the welfare of animals. Or they sarcastically pose as broccoli rights advocates. Or they start bragging aggressively about how much they love bacon.

M: Well, when you call people immoral, you have to expect them to retaliate.

V: I wouldn't. When you meet a devout Catholic, do you start going on about how much you love abortion, and how you hope to perform a few abortions later that day?

13 Lomasky ("Is it Wrong to Eat Animals?" *op. cit.*, p. 199) warns of the danger of "excessive self-satisfaction" for ethical vegetarians.

M: Well, I'm pro-choice, but I still wouldn't do that. That would be offensive.

V: Would you tell them how Catholics are just motivated by the desire to feel superior to others, or to tell others what to do?

M: No. I assume that Catholics are motivated by their understanding of God, and of what God requires of us.

V: Yet many meat-eaters refuse to accept that vegans are motivated by their understanding of what morality requires of us.

(u) Why promote veganism to others?

M: So why do it? Why not just keep your veganism to yourself?

V: (*sigh*) Because that would be wrong.

M: I don't see why. You're not making them eat meat. It's not your job to ensure other people's morality. Like how you don't have to go around making sure your neighbors aren't cheating on their spouses. Each of us is only responsible for the morality of our own actions.

V: That might be true in general. But I think this issue is a special case.

M: What's special about it?

V: First, for most of the wrongs of the past – slavery, colonialism, the oppression of women – the victims could and did speak up. In the present case, the victims will never be able to act or speak for themselves. There is no one to speak against what we humans are doing, except us. So we have to do it. If we don't, it will never stop.

M: I don't think I want to do that. It's going to be really awkward if every time I eat a meal with someone, I start telling them that they're immoral.

V: Let me give you another analogy. Remember the My Lai Massacre?

M: Some kind of mass killing in Asia, right?

V: Yeah, during the Vietnam war. American soldiers wiped out an entire village of Vietnamese civilians.[14]

M: That's awful.

V: Say you're one of the soldiers. Some of your colleagues are shooting villagers and dumping them in a mass grave. The villagers are completely unarmed. They can't even plead for their lives, because they don't speak English. What do you do?

M: Well, I wouldn't shoot the villagers.

V: In fact, some soldiers declined to participate, as you suggest. But that wasn't good enough. Then the rest of the soldiers just went on killing.

M: Well, what would you suggest?

V: There's no one else there besides the soldiers and the villagers. The villagers can't do anything to stop the massacre. So it has to be on the soldiers. The ones who know that it's wrong have to try to stop it.

M: But how could I stop it?

V: I don't know. But I think, at a minimum, you should tell the other soldiers that they're committing a war crime and that they have to stop.

M: I doubt that would work. They'd probably just keep doing it anyway.

V: Maybe so. But you should at least try. You shouldn't just stand by and watch because, "Oh, it's going to be socially awkward if I point out that my buddies are committing a war crime." Or because, "People are going to think I'm judgmental."

14 See History.com, "My Lai Massacre," 2009, http://www.history.com/topics/vietnam-war/my-lai-massacre.

M: But if I make people really uncomfortable, they're not going to stop eating meat; they'll just stop talking to me.

V: Fair point. So you need to strike a balance between staying silent in the face of a great wrong, and alienating most other people. You need to make the point that meat-eating is wrong, but in a calm and rational manner, so you don't sound crazy.

(v) How wrong is meat-eating?

M: Speaking of crazy, don't you think some of your analogies are a little overblown? Comparing meat-eating to being a Nazi, or a slave-master, or massacring a village of civilians? I don't claim to be a saint or anything, but I'm hardly a Nazi just because I enjoy the occasional burger.

V: Many people feel that way. That's another reason why people keep eating meat, even after they know that it's wrong.

M: What, because we're not saints?

V: Nor do we want to be. Most people have a certain tolerance for immorality in themselves. They don't aspire to be ideal; they just want to be *not too bad*, morally speaking.

M: Yeah, you know, I'm only human. Like, sometimes I stretch the truth when I'm trying to impress a date. I know it's wrong, but I'm not trying to be perfect. Anyway, I make up for it by being very nice.

V: Yeah, this isn't like that. The arguments we've been discussing don't suggest that meat-eating is a minor foible, like lying to impress your date. The arguments suggest that human meat consumption, taken as a whole, may be literally the greatest problem in the world.

M: Yeah, I know. 74 billion animals and all. But any given person only contributes a tiny amount of that harm.

V: True, you contribute only a tiny *percentage* of it. But that tiny percentage is still extremely harmful. If you keep eating meat, you're probably going to eat over two thousand land animals in your lifetime.[15]

M: Two thousand? I don't believe that.

V: Well, if you eat meat at every meal, three times a day, that's three times 365 – that's over 1,000 meat dishes a year, right?

M: Right . . .

V: The average American lives 79 years. So that's over 79,000 meat dishes in a lifetime. How many animals do you think would be needed to provide 79,000 meat dishes?

M: But I don't eat meat at every meal.

V: Okay. What if you do it half the time? Then you eat about 40,000 meat dishes in a lifetime. It's still plausible that over 2,000 animals would have to be killed to provide that.

M: I thought your argument was all about causing suffering, not about *killing*.

V: Yeah, I'm just using the numbers killed as a proxy for the number of animals that are subjected to extreme suffering. Subjecting more than *two thousand* other creatures to agony would be the worst thing you did in your life, by far.

(w) Comparing meat-eaters to Nazis

M: Alright, but it's still not like being a Nazi.

15 Geiss ("How Many Animals Do We Eat?" *op. cit.*) estimates that Americans eat 2,400 land animals per lifetime, based on the annual meat consumption in America, the national population, and the average life expectancy at birth. Jonathan Foer estimates that an average American eats 21,000 animals (presumably including seafood) in a lifetime (*Eating Animals*, New York, NY: Little, Brown, & Co., 2009, ebook location 1441).

V: No? Why not?

M: Well, everyone *knows* that killing people is wrong. Most of us don't know that eating meat is wrong.

V: The Nazis didn't know that killing Jews was wrong either. Many of them thought that it was morally right. Some of them talked about steeling themselves to do their duty.[16]

M: Okay, but everyone *should* know that it's wrong to murder people.

V: Everyone should also know that it's wrong to inflict severe pain and suffering for trivial reasons.

M: Okay, but not everyone knows that the meat industry does that.

V: And most Nazis didn't know, when they first became Nazis, or when they helped ship people to concentration camps, that Hitler was going to order everyone in the camps to be executed.

M: Hey, are you defending Nazis now?

V: That wasn't my intention. Nevertheless, the people we see as villains are rarely as diabolical as we imagine them. And the people we see as decent, including ourselves, are rarely as decent as we imagine them. The gap between the Nazi officer and the average American is much smaller than we think. Both are, for the most part, simply going along with what those around them are saying and doing, without thinking too hard about what is right or wrong, or trying too hard to be good or bad.

16 Jonathan Bennett discusses this in "The Conscience of Huckleberry Finn," *Philosophy* 49 (1974): 123–34, at pp. 127–9. See also Claudia Koonz, *The Nazi Conscience* (Cambridge, MA: Belknap Press, 2005).

One just happens to live in a society in which a genocide is being carried out.

M: That's pretty intense. I'm going to need to think more about all this before changing my lifestyle.

V: Fair enough. But don't put it off for too long. "The sad truth is that most evil is done by people who never made up their minds to be either good or evil."[17]

17 Here V slightly misremembers Hannah Arendt's remark: "The sad truth of the matter is that most evil is done by people who never made up their minds to be or do either evil or good" (*The Life of the Mind*, vol. 1, [San Diego, CA: Harcourt, 1978], p. 180).

Annotated Bibliography

The following is a list of sources, all of them either cited in the dialogues or directly related to the discussion. It is organized by the day of the dialogue. The list is not meant to be exhaustive, but it does include the most important publications related to ethics of vegetarianism and veganism.

Day 1

Cheryl Abbate, "Comparing Lives and Epistemic Limitations: A Critique of Regan's Lifeboat from an Unprivileged Position," *Ethics and the Environment* 20(1) (2015): 1–21.

> Philosopher Cheryl Abbate challenges the view that because humans allegedly have "higher satisfactions" than do other animals, the lives of humans have more value than the lives of nonhuman animals. She argues that this account of value has problematic implications for *humans*, such as oppressed persons who often lack the opportunity to enjoy higher satisfactions. She thus recommends that we abandon the idea that we can make comparable judgments about the value of the lives of sentient beings.

Sahar Akhatar, "Animal Pain and Welfare: Can Pain Sometimes Be Worse for Them Than for Us?" pp. 495–518 in *The Oxford Handbook of Animal Ethics*, edited by Tom L. Beauchamp and R. G. Frey (Oxford, UK: Oxford University Press, 2011).

Philosopher Sahar Akhatar challenges the common view that because nonhuman animals are less cognitively sophisticated than humans, pain for nonhumans is not as bad as pain is for humans. Because humans can, to some extent, manage their pain through expectations, memories, and so forth, Akhatar argues that for at least some nonhuman animals, a given amount of pain is worse for them than a similar amount of pain is for humans.

American Academy of Nutrition and Dietetics, "Position of the American Dietetic Association: Vegetarian Diets," *Journal of the American Dietetic Association* 109(7) (2009): 1266–82.

The American Academy of Nutrition and Dietetics, formerly the American Dietetic Association, is the world's largest organization of food and nutrition professionals devoted exclusively to nutrition and dietetics. In this position paper, they conclude that appropriately planned plant-based diets are not only nutritionally adequate, but may also prevent certain diseases, such as cancer and type 2 diabetes.

Peter Carruthers, *The Animals Issue: Morality in Practice* (Cambridge, UK: Cambridge University Press, 1992).

Philosopher Peter Carruthers defends a version of contractualism, a theory which typically holds that only rational beings are entitled to direct moral consideration and thus excludes both marginal humans and nonhuman animals from direct moral concern. Yet, Carruthers argues that humans who are not moral agents should be accorded full moral status because failing to do so is a "slippery slope" to the mistreatment of rational humans.

David DeGrazia, *Taking Animals Seriously: Mental Life and Moral Status* (Cambridge, UK: Cambridge University Press, 1996).

Philosopher David DeGrazia provides a thorough discussion of whether the interests of nonhuman animals should be afforded equal consideration. He moreover provides a detailed account of the mental lives of nonhumans, drawing attention to their feelings, beliefs, desires, language capacities, and self-awareness.

Frans de Waal, *Are We Smart Enough to Know How Smart Animals Are?* (New York, NY: W. W. Norton & Company, 2017).

This book explores the complexities of nonhuman animal cognition and intelligence. De Waal, a primatologist and ethologist, reflects upon the complex cognitive abilities of a wide variety of animals, including crows, dolphins, parrots, sheep, wasps, bats, chimpanzees, and bonobos, highlighting their use of tools, cooperation, awareness of individual identity, theory of mind, planning, metacognition and perceptions of time.

Joan Dunayer, *Speciesism* (Derwood, MD: Ryce Publishing, 2004).

Animal rights advocate Joan Dunayer charges mainstream animal liberation theories with speciesism because they do not defend a strictly egalitarian approach to animal ethics. Because some animal rights theorists, such as Tom Regan, believe that we ought to prioritize humans over nonhuman animals in forced-choice situations, Dunayer concludes that even animal rights theorists are speciesist (or as she puts it "new-speciesists") insofar as they fail to accord nonhuman animals the equal consideration and respect they are due.

Jonathan Safran Foer, *Eating Animals* (New York, NY: Little, Brown, and Company, 2009).

Novelist Jonathan Safran Foer exposes the cruelty of industrialized animal farming, in part by drawing on what he himself witnessed while sneaking onto a factory farm. Although he does not argue that eating meat is intrinsically bad, Foer argues that eating industrial meat products is bad, given the meat industry's negative impacts on animals, human health, and the environment.

Andrew Linzey, *Why Animal Suffering Matters* (Oxford, UK: Oxford University Press, 2009).

Responding to those who claim that humans are superior to other animals because humans allegedly are so different, animal theologian Andrew Linzey makes the case that these so-called differences between humans and nonhuman animals actually give humans all the more reason to foster a moral concern for other animals. For instance, because most nonhuman animals lack reason, they cannot consent to the harm that we cause them, and because they lack a soul, they will never be compensated for this harm.

Alastair Norcross, "Puppies, Pigs, and People: Eating Meat and Marginal Cases," *Philosophical Perspectives* 18(1) (2004): 229–45.

Philosopher Alastair Norcross argues that consuming factory-raised meat is just as bad as torturing puppies for pleasure because both cause serious harm to nonhuman animals for trivial pleasure. Norcross acknowledges that, since one individual cannot, by herself, influence the market, there is a chance that the behavior of an individual meat-eater is not harmful. Yet, he goes on to argue that even so, it's wrong to consume factory raised products because there is a risk that the behavior of meat-eaters *is* harmful, and even small risks of great harm are morally unacceptable.

Evelyn Pluhar, *Beyond Prejudice: The Moral Significance of Human and Nonhuman Animals* (London, UK: Duke University Press, 1995).

Philosopher Evelyn Pluhar challenges the view that only full-fledged persons have moral rights. Such a view, she argues, has problematic implications for humans who aren't persons, such as the severely mentally disabled. Her view is that sentience is what grounds moral status, which implies that sentient non-human animals are just as morally significant as full-fledged human persons.

James Rachels, *Created from Animals: The Moral Implications of Darwinism* (Oxford, UK: Oxford University Press, 1990).

Philosopher James Rachels shows how Darwin's theory of evolution undermines the idea that while humans have special moral worth, nonhumans have little value. Rachels defends "moral individualism," the idea that the value of a being depends on the being's individual capacities and characteristics, and not the being's species membership. On his view, there is a sense in which humans are "devalued," while the value of nonhuman animals is increased.

Tom Regan, *The Case for Animal Rights*, 2nd ed. (Berkeley, CA: University of California Press, 2004).

In this seminal book, philosopher Tom Regan defends the view that nonhuman animals have rights and that these rights are equal to the rights of humans. He, however, argues that although humans and nonhuman animals have the same kind of value (inherent value), the value of their lives are different. Because humans allegedly have greater opportunities for satisfaction, such as the satisfaction of acting on abstract moral principles, their *lives* have more value. While this doesn't mean that we can then exploit nonhuman animals for human gain, it does, according to Regan, imply that we ought to save rational humans over nonhumans in forced-choice situations.

Dario Ringach, "The Use of Nonhuman Animals in Biomedical Research," *The American Journal of the Medical Sciences* 342(4) (2011): 305–13.

Dario Ringach, a Professor of Neurobiology and Psychology, outspoken critic of animal rights and defender of speciesism, argues that it's acceptable to use nonhuman animals as resources because he believes that the lives of humans are more valuable than the lives of nonhumans. He makes the case that there is unequal moral status between humans and other animals by pointing out that if we were forced to choose between saving a mouse or saving a human in a fire, our intuitions would inform us that we ought to save the human. Because of this, Ringach concludes that humans have more value than other animals, and thus it's permissible to use nonhumans as resources.

Peter Singer, *Animal Liberation* (New York, NY: HarperCollins, 2009).

This seminal book by philosopher Peter Singer defends the basic principle of equality, which entails that we ought to give equal consideration to the interests of all sentient beings. Singer applies this principle to the ethics of factory farming, and after providing lengthy and detailed descriptions of the cruelties of factory farming, he concludes that it's morally wrong to consume factory-farmed products.

Craig Stanford, *The New Chimpanzee: A Twenty-First-Century Portrait of Our Closest Kin* (Cambridge, MA: Harvard University Press, 2018).

Biologist and anthropologist Craig Stanford reviews the past two decades of chimpanzee research, highlighting chimpanzee social behavior, politics, and communication. This book draws attention to the kinship great apes have with humans, helping us to better understand what it means to be human.

P. J. Tuso, M. H. Ismail, B. P. Ha, C. Bartolotto, "Nutritional Update for Physicians: Plant-Based Diets," *The Permanente Journal* 17(2) (2013): 61–6.

In this paper, medical experts advise physicians to recommend a plant-based diet to patients, especially those with high blood pressure, cardiovascular disease, diabetes, or obesity. As they argue, a plant-based diet is the best way to achieve healthy eating.

Day 2

Mark Budolfson, "Is It Wrong to Eat Meat from Factory Farms? If So, Why?" pp. 30–47 in *The Moral Complexities of Eating Meat*, edited by Ben Bramble and Bob Fischer (New York, NY: Oxford University Press).

The "inefficacy objection" to the common arguments for vegetarianism suggests that buying meat from the supermarket is like dumpster diving because an individual's consumption of animal products allegedly does not impact animal suffering. To explain why eating factory-farmed animals is nevertheless objectionable, philosopher Mark Budolfson invokes the notion of *the degree of essentiality of harm to an act* and argues that consuming factory-farmed products is particularly objectionable because it is essential that harm is tied to its production.

Carl Cohen, "The Case for the Use of Animals in Biomedical Research," *The New England Journal of Medicine* 315(14) (1986): 865–70.

Philosopher Carl Cohen argues that in order to have rights, one must have the capacity for what he calls "free moral judgment," and he claims that humans have, while nonhumans lack, this capacity. While he acknowledges that some humans (often referred to as "marginal human beings") lack this capacity, Cohen argues that even these humans have rights because they are of a kind that possesses the capacity for free moral judgment.

Hud Hudson, "Collective Responsibility and Moral Vegetarianism," *Journal of Social Philosophy* 24(2) (1993): 89–104.

Philosopher Hud Hudson argues that those who want to defend moral vegetarianism ought to be attentive to theories of collective responsibility. Because the use of animals for food is essentially a collective action problem, Hudson appeals to the notions of collective responsibility and collective inaction to explain why individuals have a moral reason to become vegetarian.

Andrew Linzey, *Animal Theology* (London, UK: SCK Press, 1994).

Animal theologian Andrew Linzey bridges the gap between animal rights and theology by providing a theological grounding for the rights of animals. He argues that, contrary to common thought, the biblical notion of "dominion" does not mean "despotism," but rather "stewardship," and being good stewards of the earth requires that we treat nonhuman animals with respect.

Dan Lowe, "Common Arguments for the Moral Acceptability of Eating Meat," *Between the Species* 19(1) (2016): 172–92.

Philosopher Dan Lowe considers and responds to three common defenses of meat-eating, including the argument that eating meat is permissible because it's "natural." Lowe points out that what's natural is not always good, and thus the "naturalness" argument has counterintuitive implications.

Nathan Nobis, "Vegetarianism and Virtue: Does Consequentialism Demand Too Little?" *Social Theory and Practice* 28(1) (2002): 135–56.

Philosopher Nathan Nobis brings together consequentialism and virtue theory to make a case that individuals ought to be vegetarian, even if an individual's act of consuming meat does not cause harm. As he argues, those who foster the virtue of compassion are more likely to bring about goodness in the world, and since a compassionate person would not derive pleasure from consuming animal products, knowing that animals suffered immensely and died for the products, individuals ought to become vegetarian.

Evelyn Pluhar, "Moral Agents and Moral Patients," *Between the Species* 4(1) (1998): 32–45.

Philosopher Evelyn Pluhar emphasizes the moral distinction between "moral patients" and "moral agents." She suggests candidates for both "moral agency" and "moral patiency" and considers whether nonhuman animals fall into either category.

Russ Shafer-Landau, "Vegetarianism, Causation and Ethical Theory," *Public Affairs Quarterly* 8(1) (1994): 85–100.

The "causal impotence" objection to vegetarianism claims that even if factory farming is wrong, it's not clear that individuals have a duty to become vegetarian. After all, one individual's decision to consume meat does not by itself cause harm. Philosopher Russ Shafer-Landau counters that even if an individual's act of meat-eating does not cause harm, there is still reason to adopt a vegetarian diet. As he suggests, meat-eaters seem to display an indifference to the cruelties involved in the production of their food, and for this, they may be condemnable.

Matthew Scully, *Dominion* (New York, NY: St. Martin's Griffin, 2002).

Conservative journalist Matthew Scully challenges the claim that the Bible permits humans to use other animals in any way they see fit. Because we are more powerful than nonhuman animals, and because they are unequal to us, we are responsible for treating them with kindness and empathy. To properly exercise our dominion, we should act as decent, faithful stewards of the Earth and have mercy for the creatures who live in it – human and nonhuman.

Day 3

Elise Aaltola, "Politico-Moral Apathy and Omnivore's Akrasia: Views from the Rationalist Tradition," *Politics and Animals* 1 (2015): 35–49.

Philosopher Elise Aaltola draws attention to the problem of what she calls "omnivore's akrasia," a condition in which one rationally recognizes that one ought not to consume animals, but continues to eat them. She suggests that our society and culture is, in part, responsible for omnivore's akrasia. As an external influence on individual consumers, consumerist society causes individuals to foster negative emotions or apathy toward farmed animals.

Colin Allen, "Animal Pain," *Noûs* 38(4) (2004): 617–43.

Philosopher Colin Allen examines recent philosophical arguments for and against nonhuman animal consciousness and draws on scientific research to challenge the main arguments. His article is a call for philosophers to consider seriously the scientific research on nonhuman animal minds, which is needed to inform discussions of nonhuman consciousness.

Marc Bekoff, *The Emotional Lives of Animals* (Novato, CA: New World Library, 2007).

Appealing to Charles Darwin's principle of evolutionary continuity, cognitive ethologist Marc Bekoff draws attention to the rich and deep emotional lives of nonhuman animals. Sharing stories of animal joy, empathy, grief, embarrassment, anger, and love, Bekoff explains how animal emotions evolved to bond social animals with one another.

Peter Carruthers, "Brute Experience," *Journal of Philosophy* 86 (1989): 258–69.

Philosopher Peter Carruthers' "higher-order thought" theory of consciousness leads him to conclude that the experiences of other animals, including pain, are nonconscious and therefore not morally concerning. On his view, nonhuman animal behavior is on par with nonconscious human behavior, such as driving while distracted.

Melanie Joy, *Why We Love Dogs, Eat Pigs, and Wear Cows: An Introduction to Carnism* (San Francisco, CA: Conari Press, 2010).

Social psychologist Melanie Joy discusses what she calls "carnism"– the cultural belief system that normalizes meat-eating and causes people to disconnect from their natural tendencies to be compassionate and empathic. After responding to and rejecting the common arguments that meat-eating is normal, natural, and necessary, she concludes that eating animals is always unethical.

Norton Nelkin, "Pains and Pain Sensations," *Journal of Philosophy* 83 (1986): 129–47.

Philosopher Norton Nelkin explores whether nonhuman animals are conscious and what their consciousness is like. While those who defend the view that other animals are conscious (and that their consciousness is like ours) often point to the similarities between humans and nonhumans, Nelkin points out that the differences between nonhumans and humans suggests that the sensations of other animals may be different from human sensations.

Jesse Prinz, "Is Empathy Necessary for Morality?" pp. 211–29 in *Empathy: Philosophical and Psychological Perspectives*, edited by Amy Coplan and Peter Goldie (Oxford, UK: Oxford University Press, 2011).

Philosophy professor Jesse Prinz argues that empathy, which involves feeling, to some degree, what another feels, is biased insofar as we tend to feel greater empathy for those who are similar to ourselves. He moreover contends that empathy is vulnerable to "cuteness effect bias," which may explain why many humans feel empathy for cats and dogs, but not farmed animals.

Stuart Rachels, "Vegetarianism," pp. 877–905 in *The Oxford Handbook of Animal Ethics*, edited by Tom L. Beauchamp and R. G. Frey (Oxford, UK: Oxford University Press, 2011).

Philosopher Stuart Rachels provides a detailed overview of the harms of industrial farming, focusing on not only cruelty to animals, but also the harms factory farms cause to the environment, rural communities, and human health. He controversially argues that the intensity and extent of farmed animal suffering is so great that industrial farming has caused (at least) five thousand times more intense pain than the Holocaust.

Bernard Rollin, *The Unheeded Cry: Animal Consciousness, Animal Pain and Science* (New York, NY: Oxford University Press, 1988).

Philosopher Bernard Rollin provides detailed evidence for the view that nonhuman animals are conscious and experience pain. He moreover explores the origin of the view that denies consciousness to animals – a view that is often accepted by scientists themselves – and he calls for scientists to acknowledge the scientifically informed view that animals are indeed conscious and experience the pains inflicted upon them.

John Searle, "Animal Minds," *Midwest Studies in Philosophy* 19 (1994): 206–19.

Philosopher John Searle argues that at least some nonhuman animals are conscious, act intentionally, and have thought processes. Searle judges that, through our interactions with other animals, we can infer, and moreover *know*, that some nonhumans are conscious. Because Searle believes that it is obvious that some nonhuman animals are conscious, his essay is more of a response to those who deny that other animals are conscious, and not a positive argument for nonhuman consciousness.

Marjorie Spiegel, *The Dreaded Comparison: Human and Animal Slavery* (New York, NY: Mirror Books, 1997).

Marjorie Spiegel, executive director of the Institute for the Development of Earth Awareness, draws attention to the similarities between the violence humans have committed against other humans, especially through slavery, and the violence humans currently inflict upon other animals. She claims that the similarities between the oppression of humans and the oppression of animals ought to be acknowledged so we can better understand the fundamental cause of both individual-level and societal violence.

Gary Varner, *In Nature's Interests? Interests, Animal Rights, and Environmental Ethics* (New York, NY: Oxford University Press, 1998).

Philosopher Gary Varner advances a theory of animal consciousness which holds that there are six conditions that are relevant to conscious experience of pain. He claims that while there is evidence that mammals meet each of the six conditions and that birds meet five conditions, the so-called lack of evidence for invertebrate sentience suggests that we ought to draw the "sentience line" between vertebrates and invertebrates.

Day 4

Cheryl Abbate "Veganism, (Almost) Harm-Free Animal Flesh, and Nonmaleficence: Navigating Dietary Ethics in an Unjust World" in *Routledge Handbook of Animal Ethics*, edited by Bob Fischer (New York, NY: Routledge, 2019).

While philosopher Cheryl Abbate argues that it's impermissible to consume sentient animals raised on either factory farms or "humane" farms, she defends ostroveganism, which holds that it's permissible to eat the flesh of nonsentient animals, such as bivalves. Vegans who oppose the consumption of bivalves are guilty of what she calls "kingdomism" – the view that an animal is entitled to serious moral consideration just because of its membership in the animal kingdom.

David DeGrazia, "Moral Vegetarianism from a Very Broad Basis," *Journal of Moral Philosophy* 6 (2009): 143–65.

Philosopher David DeGrazia provides an argument for vegetarianism that assumes neither that nonhuman animals have moral rights nor that we should give them equal consideration. His argument stems from the very weak claim that nonhuman animals have moral status, and this, he argues, is enough reason to become vegetarian.

Dan Demetriou and Bob Fischer, "Dignitarian Hunting: A Rights-based Defense," *Social Theory and Practice* 44(1) (2018): 49–73.

Philosophers Dan Demetriou and Bob Fischer point out that many field animals are killed in the process of plowing, planting, harvesting, protecting crops, and pest management, thus even industrial plant agriculture causes great harm to nonhuman animals. As an alternative to veganism, they propose that consumers adopt the philosophy of "dignitarian" hunting, which they claim is more in line with animal-rights theory than the consumption of a strictly plant-based diet.

Bob Fischer, "Bugging the Strict Vegan," *Journal of Agricultural and Environmental Ethics* 29 (2016): 255–63.

Philosopher Bob Fischer argues that if we have a choice between harming beings that we know to be sentient or harming beings that we don't know to be sentient, we should choose the latter. Since plant production does cause harm to animals who are clearly conscious, such as field mice who are frequently run over by plow machines, he suggests that we reduce our consumption of plants by eating "maybe sentient" animals, such as insects.

Elizabeth Harman, "The Moral Significance of Animal Pain and Animal Death," pp. 726–37 in *The Oxford Handbook of Animal Ethics*, edited by Tom L. Beauchamp and R. G. Frey (Oxford, UK: Oxford University Press, 2011).

Philosopher Elizabeth Harman argues that if there are strong moral reasons not to cause pain to nonhuman animals, there also are strong moral reasons not to kill them, even painlessly. She suggests that if we lack reasons not to kill nonhuman animals, then there are only weak reasons not to cause them pain – and the latter claim is clearly counterintuitive.

Colin Klein and Andrew B. Barron, "Insects Have the Capacity for Subjective Experience," *Animal Sentience* 9(1) (2016): 2–19.

Philosopher of neuroscience Colin Klein and scientist Andrew Barron make the case that insects have the capacity for subjective experience, which is described as basic awareness of the world. This capacity for basic consciousness, they argue, is supported by the integrated midbrain and basal ganglia structures of insects.

Don Marquis, "Why Abortion Is Immoral," *Journal of Philosophy* 86(4) (1989): 183–202.

Philosopher Don Marquis argues that death harms normal humans because it deprives them of a "future like ours." This view can be used to argue that both abortion and the killing of sentient (and potentially sentient) nonhuman animals are prima facie seriously wrong. Since fetuses and sentient nonhuman animals have future experiences, activities, projects, and enjoyments, death harms them insofar as it deprives them of these things.

Chris Meyers, "Why it Is Morally Good to Eat (Certain Kinds of) Meat: The Case for Entomophagy," *Southwest Philosophy Review* 29(1) (2012): 119–26.

Chris Meyers argues that because it is unlikely that insects are conscious, we should move toward entomophagy – an insect based diet, which is relatively sustainable. His claim is not just that it's acceptable to eat insects; rather, he claims that it is morally good to do so, given that an insect-based diet causes less harm than a strictly plant-based diet.

Alastair Norcross, "The Significance of Death for Animals," pp. 465–74 in *The Oxford Handbook of Philosophy of Death*, edited by Ben Bradley, Fred Feldman, and Jens Johansson (Oxford, UK: Oxford University Press, 2012).

Philosopher Alastair Norcross argues that the death of any sentient animal has moral significance. Because sentient nonhumans often experience enjoyment, their death prevents the existence of well-being. However, as Norcross notes, this does not entail that death has significance *to nonhuman animals themselves*.

James Rachels, "The Basic Arguments for Vegetarianism," pp. 70–80 in *Food for Thought: The Debate over Eating Meat*, edited by S. F. Sapontzis (Amherst, NY: Prometheus Books, 2004).

Philosopher James Rachels advances what he calls the basic argument for vegetarianism, which is supported by a simple principle that every decent person accepts: it is wrong to cause pain unless there is a good enough reason. With this basic argument, Rachels illustrates that one can argue for vegetarianism without appealing to any particular ethical theory.

Tom Regan, "The Case for Animal Rights," pp. 13–26 in *In Defense of Animals*, edited by Peter Singer (New York, NY: Basil Blackwell, 1985).

This paper, which is a sketch of philosopher Tom Regan's seminal book *The Case for Animal Rights*, argues that nonhuman animals have the fundamental right not to be viewed or treated as our resources. Consequently, it is wrong to use them for food, even if they are treated "nicely" before they are killed. On Regan's view, giving farmed animals "happy lives" will not touch the basic wrong of viewing and treating them as our resources.

Andrew Smith, *A Critique of the Moral Defense of Vegetarianism* (New York, NY: Palgrave Macmillan, 2016).

Philosopher Andrew Smith argues that plant neurobiology provides us with enough scientific evidence to conclude that plants are sentient, and thus we need a new defense of vegetarianism to accommodate the possibility of plant sentience. Smith defends "expansionary sentientism," which is the position that plants have equal moral standing because they, too, are sentient.

Vasile Stănescu, "Beyond Happy Meat: The (Im)Possibilities of 'Humane,' 'Local,' and 'Compassionate' Meat," pp. 133–55 in *The Future of Meat without Animals*, edited by Brianna Donaldson and Christopher Carter (London, UK: Rowman & Littlefield, 2016).

Professor of Communication Vasile Stănescu criticizes the "compassionate carnivore" movement, pointing out that even "humane" meat causes unnecessary harm to animals and is environmentally destructive. This chapter gives an in-depth review of the harms animals endure on "happy farms" and the environmental destruction that would be caused if billions of animals were raised on "free-range" farms.

Tatjana Višak, *Killing Happy Animals: Explorations in Utilitarian Ethics* (New York, NY: Palgrave Macmillan, 2013).

Philosopher Tatjana Višak addresses whether it is permissible to kill a healthy nonhuman animal who was provided with a happy life, as is allegedly done in the production of "humane" meat. In answering this question, she considers two different versions of utilitarianism and their implications for killing happy nonhuman animals.

Corey Wrenn, "Nonhuman Animal Rights, Alternative Food Systems, and the Non-Profit Industrial Complex," *Phaenex: Journal of Existential and Phenomenological Theory and Culture* 8(2) (2013): 209–42.

Sociologist Corey Wrenn criticizes "humane" farms because, as she argues, any food system that uses nonhuman animals is inherently problematic. All animal agriculture systems perpetuate the idea that nonhuman animals are property, and this ultimately undermines any serious consideration for their well-being.

Index